The Comfort Demands

Occupy Pharmacy

By: Jim Plagakis, RPh

Copyright by Jim Plagakis 2011

ISBN: 978-1-105-34659-0

"Power Concedes Nothing Without Demands"

Frederick Douglass

How Can you Get Where You Want to Go

If you Don't Know Where You Are?

Or How you got there?

Forward

This is the logical conclusion of the "Comfort" messages. It follows "The Comfort Trilogy" and details how we got to the soul-crushing conditions that predominate in retail pharmacy in the 21st Century.

In this work, I will give a history of pharmacy from my view of having worked in drug stores/pharmacies for 54 years. In 1957, I took my first drug store job. I was a stock boy. In August of 1964, I became registered in Ohio. I have done nothing else.

I have observed the progression of the job of working in a pharmacy as it evolved through three phases. I will call them *The Historical Era, The Modern Era and The Dark Era.* Seeing how we got to where we are will help us to move on.

I will conclude this book by outlining reasonable and uncomplicated demands that will benefit both the institutionalized pharmacist and the institution that suffers right along with us. A happy pharmacist is a more productive employee and that means more profits.

Please forgive me for the redundancies. I repeat things I think are really important. That is just what I do. When a professional editor gets involved in future editions, that will change and this will be a collector's item.

Jim Plagakis, January, 2012

If you don't know what you want,
*First find out what you **don't want.***

How Did it Ever get To Be this bad?

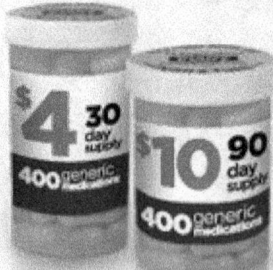

At this point, I honestly don't care what price they sell prescriptions for. They can give them away (and they do). All I want is to make a good living and the opportunity to practice pharmacy, fulfill my standards and follow the law. The one problem I see with the $4.00 or free prescriptions is that they detract from the patients' perceptions that we provide a valuable professional service.

As long as we are tied to the product, price will be an important factor. Americans believe that the more expensive a product is, the better it is. To most people getting valuable professional service when the price is $4.00 is not believable.

The Veneer

Way back in the day, great works of art were protected by a layer of shellac. Oil paints were made from natural materials including egg whites. Left all alone on the canvas, the colors would fade. In humidity, they could get moldy or be infested by bacteria. This layer of varnish invariably diminished the beauty of the paintings. It could get yellowy or even brown. Modern experts are paid a lot of money to remove the veneer without harming the painting.

Veneer: An outer layer applied to a surface. A superficial appearance or show put on to please others.

There is a veneer covering the practice of pharmacy. This veneer is the processes and procedures that we use to prepare the final product. This used to be called "Filling Prescriptions" and it was the be all, end all of pharmacy practice. Filling prescriptions was our professional task. It was what we did. Filling prescriptions was the extent of our professional lives. We did not counsel openly because doctors accused us of interfering with the doctor-patient relationship. Every single new pharmacist accepted this role and it is carrying over into the 21st Century.

I am astounded when I listen to pharmacists talk about the end of pharmacy. They continually see moves to streamline prescription filling by companies like Walgreens as not so secret long term plans to eliminate pharmacists from their long time role as prescription-fillers. They are afraid that pharmacists will be laid off right and left. They see it as a bad thing that non-pharmacists will be doing the prescription filling.

In 1971, I attended a meeting of the California State Board of Pharmacy. The subject of allowing non-pharmacists to assist in the filling of prescriptions was on the agenda. The board was considering an official designation for that non-pharmacist: *Pharmacy Technician.*

The union was there. The Guild was there. Individual pharmacists were there. They were having an official hissy fit about this. They said that non-pharmacists would push pharmacists out on the street, that there would be layoffs of pharmacists. I was the only one to speak in favor of technicians. You now know how that turned out. We have technicians and we could not so the job without them. I predict that the same will happen again, soon .

I do not see competent pharmacists losing their jobs. For years now, we have put up with pharmacists we called *Warm Bodies With Licenses* behind the counter. We have paid them a full pharmacist's wage. We have cringed when we viewed their questionable competence. We put up with them because we wanted our days off.

There was a time during the late 1980s when I had to hire a *Warm Body.* I had no choice. The pharmacy was in a Pay 'n Save drug store on Whidbey Island, north of Seattle. No pharmacist from the Interstate-5 corridor wanted to move there. There was a pharmacist shortage in

Seattle. Why should they come to a place where there was nothing for a young person to do? The closest town on the mainland was Mount Vernon. Depending on the weather and the time of year, the commute could be 45 minutes to an hour and a half. Deception Pass State Park is the most popular park in the state of Washington. During the summer, there was always a traffic jam at the Deception Pass Bridge. As we streamline, *Warm Bodies with Licenses* will probably not find steady jobs.

This pharmacist was barely competent. He had difficulties with the computer. To get by, he often would run *labels only* from the last filling to speed things up. He might sell 100 prescriptions and the log would show 60. It took forever to get rid of him because nobody wanted the job. There still may be a shortage of pharmacists in the 21st Century in outlying areas. However, with the puny reimbursements from the PBMs, I don't see any big company doubling up on pharmacists. We will have to see how the state legislatures bend the laws to help us streamline the prescription filling process.

I do not believe that pharmacists who are challenged with the English language and unfamiliar with

American life will have jobs. The days of sponsoring H1b Visa holders will be coming to an end. These people are not on a citizenship track. They are here to make money. Their American annual pharmacist's wage can support their entire family for years in Bangladesh. The companies that have sponsored H1b Visa holders basically get indentured servants, but an American pharmacist, even a kid wet behind the ears, is much more appealing to an American baby boomer.

I have an Abdul story. This guy worked for CVS. He called for a copy. There was evidence that he was clearly not prepared to practice pharmacy in the United States of America. There were no refills on the prescription.

Abdul said, "You will call the doctor for refills and then call me back?" He had to say it three times because his accent was so thick that I couldn't believe what he said.

"You want me to what?" First he had kept me on hold and now this?

"Since you have the prescription on file, I want you to call the doctor before the transfer."

"Not going to happen, Abdul. How long have you been practicing pharmacy in the United States?"

"I have been here for four months."

"This is not Egypt, Abdul."

"I am not from Egypt."

"You need to talk with your Pharmacy Manager and get educated on the protocols that we follow."

He got huffy, "I am the Pharmacy Manager."

Heaven Help Us

H1b Visa Sponsorship Applications
Does Not include Green Cards

2001 to 2008		2010
295	Kroger	11
359	Safeway	27
4791	CVS	499
3448	Rite-Aid	244
1477	Walgreens	27

I found it interesting and perplexing that Rite-Aid applied to sponsor one foreign pharmacist for the job of Regional Vice President in 2010.

I believe that the days of H1b pharmacists working in the United States are numbered. The sponsorship process is not cheap. For fun, visit some of the websites that discuss H1b Visa pharmacists. There are discussion boards that will amaze you (and piss many of you off). Comments by a young Filipino female pharmacists are beyond belief. She had gotten and H1b Visa job in California with CVS. She did a verbal end zone dance.

The 21st Century has begun and H1b Visa pharmacists will not be included. However, some of these companies continue to amaze me. I would not be surprised by anything.

Marginal pharmacists may have a difficult time keeping a good job. Pharmacists who are unwilling or unable to adapt to the coming changes will have difficulty. I have worked with pharmacists who seem to think that their only job will always be reviewing and verifying prescriptions and checking the work of technicians. They will be left behind.

Pharmacist with little or no business skills, Cranky pharmacists, abusive pharmacists, old and tired pharmacists, too young pharmacists, physically challenged pharmacists, fat pharmacists, skinny pharmacists, stuttering pharmacists, foreign pharmacists, pharmacist from the new boutique pharmacy schools, all may have trouble finding good jobs. This has been the way it has been in law, accounting, and other professions. But, maybe not. There is a huge rumble in the distance and that rumble is a tsunami of prescriptions.

There will be plenty of pharmacists who will choose to open their own businesses and they will do well. They know that, in the end, it is a cult of personality. They will remember patients' names. They will buy the caps and team jerseys for Little League teams. They will attend weddings and they will go to funerals. They will also understand that they need a niche business that will generate income. They know that, left unchallenged, the PBMs will continue to wreak havoc.

The smartest will invest a little money into an *Aging Well* niche business. (Contact Jim Plagakis at jprxconsulting@hotmail.com) Men who want to remain manly men search for advice. A male pharmacist who wears a badge: *Men's Health Advisor* can make a living counseling older men. There are not many older men who do not want to remain sexually active, with more muscle than fat. They want to have energy and to be able to see possibilities. For all of that, they need good health.

Female pharmacists would be wise to study *Women's Health* issues. Bone preservation, hormones, hot flashes, breast health, brain power, memory, skin, fitness and sleep are all issues you can discuss intelligently.

One area where you can make a difference is Women's Sexual problems like vaginal dryness, pain with intercourse and lack of libido. Their male doctor will give perfunctory answers. Not only are these women perplexed that what was an important part of their marriage has been ruined by age, they know that their husbands are still manly men and they are sick with worry about that younger woman at the office.

You are a woman. You are a pharmacy owner. You can study the right CE courses, surf the Internet for some answers. Stand up, Wear a badge: *Women's Health Advisor.*

You can make a good income selling supplements, but not the ones you buy from Cardinal. Buy supplements that no one else carries. A brand that is a little more expensive than Centrum. Your advice is golden. Trust yourself. First, do a Google tour of the supplements. Know what you are talking about. Baby boomers do not need calories. However, they have a higher need for micronutrients than a younger person. They cannot get the nutrients they need from their diet. Boomers must use high quality supplements.

I know a pharmacist in Vermont who does very well with Metagenics supplements. His store is in Stowe, an affluent community. This brand is probably four times the cost of regular drug store brands. His patients perceive Metagenics as special because they are presented as special. I do not know if they are four times superior to what you have on the shelf now. Metagenics does have some unique formulas. I do know that you will sell this brand if you present it properly and if your clientele has money to spend.

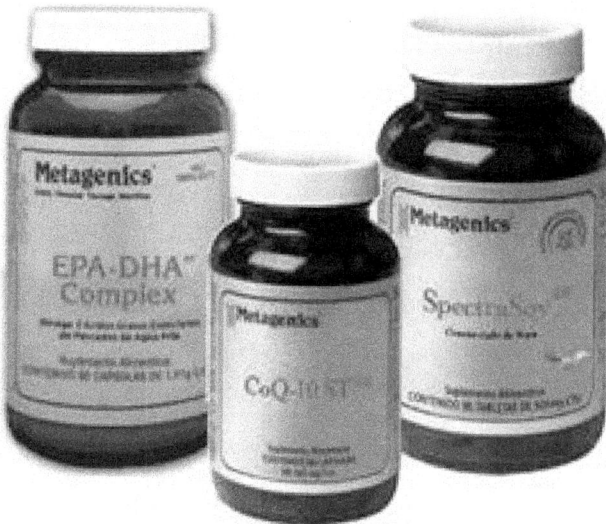

Just one example of a label that can be exclusively yours.

How will It Be Possible to Fill One Trillion Dollars Worth Of Prescriptions Every Year?

It will not be possible if we think that the present (unexamined) model of pharmacists doing the filling stays in place. If we fail, if we cannot do the job, it will be the end of pharmacy at the end of the funnel. Our culture will write us off as not up to snuff.

Dollar Amount of Prescriptions Filled 1999 compared to 2009

1999	2009
$105 Billion	$250 Billion

That is an increase $145 Billion in the decade, an increase of 138%. That is an astounding figure. In 2009, all medical care came to $2.5 Trillion. Prescriptions accounted for 10% of the total.

One Trillion Dollars by 2019? Probably

Pharmacists Will No Longer Fill Prescriptions

Pharmacists must be the ones who design the new model of technicians doing all of the duties associated with filling prescriptions

We must have *Tech Check Tech*. There must be a Technician licensure designation that certifies that the technicians qualify. We can't have the pharmacy owner's Aunt Betty's sister-in-law's granddaughter acting as an *Advanced Technician*. There must be an educational process for qualification and it must be designed by pharmacists. We are doomed if non-pharmacist business-types design the model. To allow the same MBA Masters of the Universe upper level management types that came up with *metrics* to have their hand in this would be a catastrophe.

Computer programs will have to be redone to account for the *Advanced Technician*. When the computer warns about a drug interaction (or any other problem), the process must stop cold until the pharmacist puts in his secret password to clear the interaction and allow the process to continue.

I will reluctantly concede this job of educating the new *Advanced Technicians* to the NABP and APhA. It will be slow and complicated because that is the way that the NABP and the APhA does everything, but the alternative is too worrisome. We cannot allow CVS, Rite-Aid, Wal-Mart, Kroger or Walgreens to design a course and administer that course just for their employees. This would not be acceptable. Not because Walgreens and Wal-Mart would not do a good job, but because we can take no chances that a non-pharmacist type would be involved and make choices for business reasons. The education must be standardized so any employer knows what she is getting when an applicant shows that she is licensed as an *Advanced Technician*.

An *Advanced Technician* must be an official license designation with some teeth. It must result in a real state license. You know that the CPhT designation has not gotten Certified Technicians a living wage. You all know that a CPhT designation does not even guarantee competence. These kids come out of the Junior College, they pass the test and think that they hot numbers. Then, you hire them and they fail dismally. They don't know that Methylin is locked up. They are not ready for retail.

I have worked beside CPhT technicians who did not know that the NDC is a 5-4-2 configuration. That has not been the exception. I have explained that it is *Manufacturer-Drug-Package size* and they look at me like I am speaking about Quantum Physics.

I worked with a mature man who had just become certified as a CPhT. During his first day on the job, he came to me and said, "It's time for my lunch." He had a Tupperware bowl with kind of stew in it. It smelled wonderfully gravy with big chunks of potatoes and carrots.

I said, "It's only eleven o'clock. You have only worked two hours. You need to wait until the second tech comes in at noon."

He was a man from Kenya and had a traditional upper class English accent. "I am hungry now."

"Well, tough shit, I'm hungry too. We have work to do."

"You don't have to use bad language."

"You've never heard that word before?"

"Of course I have."

"Well, man, I'm going to use it then. A tech who comes to me on the first day and tells me that he is going to lunch at the worst time pisses me off. Who do you think you are?"

"I want to have my lunch now." He had crossed his arms over his chest, stubbornly.

"Then go, but don't come back." There were two people at the counter observing this and the phone was ringing. We were at least a half hour behind because this guy was worse than slow. I had been correcting his mistakes in typing all morning.

"You aren't the manager. You can't fire me."

"No I'm not the manager, but I don't have to work with you. You are inexperienced. You are slow. You don't follow instructions. You don't seem capable of multi-tasking. You don't seem to want to learn."

He glared at me.

"I'm the pharmacist. If I refuse to work with you, how long do you think you will last?"

He went to lunch at noon and left forever about two weeks later. He said that he had expected to do *important work*, not just type and count pills.

If we do not get some laws changed to allow technicians to do almost 100% of the filling functions, with pharmacist oversight, we may not adapt as a profession fast enough. If we, as a profession, cannot efficiently and effectively fill, provide and counsel on this huge amount of prescriptions, we will be a failed profession. There could be opportunists ready and waiting to push us right out of the picture completely. I cannot ever imagine dispensaries without pharmacists, but there was a time when I could not imagine Nurse Practitioners and Physician's Assistants being the default primary medical care providers for millions of Americans.

I believe that Walgreens is separating from the other big retailers. They haven't told me their plans, but they did tell *The New York Times*.

October 21, 2011

Out From Behind the Counter

By BRUCE JAPSEN

CHICAGO —

As the <u>Walgreen</u> Company pushes its army of pharmacists into the role of medical care provider, it is bringing them out from their decades-old post behind the pharmacy counter and onto the sales floor.

The pharmacy chain, based in Deerfield, Ill., and the nation's largest, has renovated 20 stores in the Chicago area and is converting more than 40 in Indianapolis to get the pharmacist closer to patients. Pharmacists in the revamped stores are being kept away from the telephone, where dealing with insurance coverage questions and other administrative tasks occupy 25 percent of their time, Walgreen says.

"What we are seeing now is pharmacists should be using their knowledge to help consumers manage their medications appropriately," said Nimesh Jhaveri, executive director of pharmacy and health care experience at Walgreen. "It's not about the product but the care we give."

The reinvention of the pharmacist's role comes at a critical time for Walgreen, as it vies to keep its customer base. The company has so far been unable to reach a new contract with the pharmacy benefit giant, Express Scripts. At the same time, Greg Wasson, the chief executive, is trying to remake the company into a national provider of health care services.

This last summer, Walgreen sold its own pharmacy benefit management company for more than $500 million to a Maryland firm in a deal that Mr. Wasson said would help the company focus on becoming the consumer's "most convenient choice for health and daily living needs."

Walgreen braced investors last month for the potential loss next year of more than $3 billion in sales in 2012 if it lost the customers whose prescription coverage was managed by Express Scripts. In the most recent fiscal year for the company, it filled about 90 million prescriptions managed by Express Scripts. The two are parting ways effective Jan. 1 over payment issues, leaving Walgreen scrambling to contract with major employers directly in hopes that they will want to opt out of Express Scripts' pharmacy network. Walgreen's new model resembles the type of service that CVS and other major drugstore chains are trying to achieve by developing deeper relationships with customers and their doctors. Big pharmacy companies are hoping to increase reimbursements from insurers and employers as they become more integral in managing customers' medical care.

At the newly converted Walgreen stores, one of the ways pharmacists hope to develop longstanding relationships with customers is through private or semi-private consulting areas away from the busy pharmacy counter.

On Chicago's North Side, Walgreen has a pharmacy in the Andersonville neighborhood on North Clark Street that dispenses a substantial amount of medications to patients with the <u>AIDS</u> virus, so privacy for patients was critical and figured in the overall idea behind the new store model, company executives said.
Behind the pharmacy counter, the familiar bags of medications are tagged and labeled alphabetically in plastic containers, but they cannot be seen from in front of the pharmacy counter. "Customers want privacy," Mr. Jhaveri said.

The Andersonville neighborhood store includes a 50-square-foot room behind sliding doors where a pharmacist, James Wu, can sit and counsel patients, who sit on a padded bench that has enough room for the patient and a family member or two. Mr. Wu's desk is steps to the right of the private room.

Mr. Wu said he could now spend more time talking to patients or out in the store aisles, and rarely is distracted now by the orders being placed for prescriptions.
"I would take calls, asking 'Is it ready?' 'Is it covered?' " Mr. Wu said. "The phone doesn't ring anymore."
Walgreen said it would route routine questions about insurance coverage and co-payment issues to a call center in Orlando, Fla., that is staffed around the clock by

pharmacists and pharmacy technicians. Another new feature is a "health guide," a concierge of sorts who answers questions, markets new services and triages patients who may need other health care services, like treatment at a Walgreen Take Care retail clinic. At 354 of the chain's more than 7,700 stores, nurse practitioners at such clinics are available to handle routine maladies. There are financial incentives for the more personal approach; some private and government insurers have programs that reward health care providers if they can prove that their services improve the quality of care and save money.

Moreover, insurance companies and the federal government are moving to models that encourage better coordination of medical care service, putting all providers on the same page.

Federal Medicare drug laws allow for payment to pharmacists for "medication therapy management," when patients have multiple chronic diseases like hypertension, diabetes and asthma and are taking multiple medications. In recent years, Walgreen and other pharmacy chains have lobbied aggressively for reimbursement and changes to rules that allow pharmacists to do more and to get paid for these additional services. Walgreen already has aggressive lobbying efforts under way to get pharmacists the ability under state rules to administer more vaccines in the pharmacy. And the company is working with doctors and hospitals to develop relationships that include having a pharmacist involved in patient consultations and management of their diseases. "As we start to prove better outcomes, our reimbursement is going to be more based on how we do that," Mr. Jhaveri said.

Employers are open to Walgreen's idea, citing national studies showing large numbers of Americans, particularly among the elderly, who do not adhere to their treatment regimens or forget to take their medicines. For example, 2009 research from the New England Healthcare Institute showed that patients who did not take medications as prescribed cost the health system $290 billion in "avoidable medical spending every year."

"There are a variety of reasons why the current medical system is failing to help people stay on their medications," said Larry Boress, president and chief executive of the Midwest Business Group on Health, a coalition of large employers that purchases more than $3 billion in medical care services annually. Among the members are Boeing, Ford Motor and Kraft Foods.

"On filling the script, the pharmacist or pharmacy tech doesn't do much more than ask: 'Do you have any questions?' And then they give you the bag," Mr. Boress said.

Will the Other Big Players Catch On?

They better or our profession could be in big trouble. We can't have Walgreens be the only company moving to patient-centric pharmacy practice. One company cannot do it all. We can't have pharmacists working 14 hour shifts straight through and having one company contribute to putting the public in danger in thousand of stores. It is simply logical to think that a pharmacist working that long, with no breaks and no meals is dangerous. Perhaps not every day, but there will be days when she will have lost her edge. I would advise patients that they would be wise to get their prescriptions before 3:00 PM and never after 8:00 PM. The essence of the problem is that non-pharmacist MBA Masters of the Universe are captains of the ship and they do not understand that pharmacy is a profession. Pharmacies are not, by definition, dispensaries.

Americans Believe That There Is A Drug for Everything

We have been enculturated to expect drugs when we are not feeling well. Our culture demands drugs. This is not going to change. Patients are disappointed if they see the doctor for any ailment and they do not get prescriptions. Big Pharma is providing more drugs to choose from and more expensive drugs. This is not going to end. We are the last stop in the medical stream. Without pharmacy, there would be no treatment in most cases. What about that kind of power do you not understand? You are still the driver of the train. You just do not get it.

This, I repeat, we must do and we must begin now. We have to manage the processes and procedures of providing prescriptions in a manner that keeps pharmacists in charge. Our profession must be the "decider" in this picture. Otherwise, some non-pharmacist MBA types will be petitioning the state boards for changes that will not benefit the profession of pharmacy, pharmacists or patients. We allowed non-pharmacists who designed the dark ages of pharmacy to be involved in the decisions. We cannot allow them to usurp any more authority.

This is what will allow us to provide a trillion dollars worth of prescriptions in one year.

There will be more drugs and the drugs that people really want will be expensive. Big Pharma will be making more new drugs for healthy people. Viagra, Cialis and Levitra were the first big ones. Fifteen dollar tablets. The Boomers will not grow old gracefully. They want to be youthful, vibrant, strong and sexy for the whole ride and Pharma will pull out the stops to give them what they want.

Have you seen the television commercials for "Low T"? Testosterone supplementation will be very big very soon, to follow estrogen replacement therapy. This represents a brilliant opportunity for compounding pharmacists (and that is most of us). Why give this business to Big Pharma? You can compound custom made strengths. Pharma will make standardized strengths. You will have to collaborate with some doctors, but that should be easy enough. Doctors are well known for protecting their turf, but there will soon be so much turf that only an idiot would not want our help.

Starting in 2009, 10,000 Americans turn 65 years old every single day. This will continue for over 25 years. People 65 and older take an average of five prescriptions.

Forty years ago, it was around two. Mainly because there weren't that many drugs. In 1965, a high blood pressure patient took a diuretic and hydralazine. Reserpine worked well, but it despressed the patient. Two drugs, if the doctor was on the ball and many were not.

Propranolol was in clinical trials at Stanford in the early 1970s. One of my 50-something patients was a participant and was lucky enough to get the real drug. He went from grey and pasty and always winded to pink skinned and able to walk 18 holes on a regular basis. I will always remember him coming into the pharmacy with a new, my guess would be 40-something, girl friend. He handed me a prescription for Inderal. I had gotten an initial order a few days before.

"This is the drug I got at Stanford," he said, "I have a whole new life, Jim, a whole new life." He smiled, put his arm around the waist of the thin, attractive woman. "I think that I will get married again, Jim."

She leaned against him and they grinned like Cheshire cats. My respect and loyalty for Big Pharma was at a high point that day.

Big Pharma has so many drugs for high blood pressure now that some patients take three or more prescriptions. Add a statin and a proton pump inhibitor, plus the Rx Only NSAID for aches and pains and you have six prescriptions for a relatively healthy person. Add a drug for Erectile Dysfunction and one to prevent baldness and another …. Can you see the possibilities?

The Baby Boomers all by themselves are going to put enormous pressure on our industry. The population of the country is growing. We are around 310,000,000 now and should add another 50,000,000 in 10 or 15 years. They will need prescriptions. A culture that thinks there is a drug for everything will demand new drugs.

A One Sentence Recap

One Trillion Dollars
Worth of prescriptions will not be
filled by pharmacists.

The Modern Era of American Pharmacy

Before we discuss the *modern era of pharmacy*, I think we need to look at where we came from. *The Historical Era*. Until the first pharmacy computer systems, we did everything by hand. We determined prices by a set formula and used a calculator to compute the selling prices. There were no calculators until the early 1970s. They were expensive. Almost a day's wage for a basic model by Texas Instruments. Before that, we calculated price by using pencil and paper. All-Med International owned the drug store, but I designed the pricing structure for prescriptions. Here are some examples.

Top Ten Highly Competitive Tablets and Capsules

An Rx with a **net** cost of $4.00 divided by 0.92 = a selling price of $4.35 add up to $4.99

Next 20 Fast Mover Tablets & Capsules.

An Rx with a **net** cost of $4.00 divided by 0.8 = a selling price of $5.00, add up to $5.59.

Non-Competitive Tablets and Capsules for Chronic Conditions

An Rx with a **net** cost of $4.00 divided by 0.70 = selling price of $5.71, add up to $5.99

One Shot Tablets and Capsules

An Rx with an **AWP** cost of $4.00 divided by 0.6 = selling price of

Oral Liquids
An Rx for a cough syrup with an **AWP** cost of $4.00 divided by 0.5 = a selling price of $8.00, add up to $8.99.

One Shot Eye Preparations
An Rx with an **AWP** cost of $4.00 divided by 0.4 = a selling price of $10.00, Add up to $10.99.

You get the idea. If four tablets of Methergine cost **AWP** $4.00, the selling price would never be less than $12.00. During the 1970s, this was usually a one-shot post-abortion drug.

In 1972, 90% of all Prescriptions were cash.

The All-Med Drug Store that I managed was in Pacheco, California. There were a Payless NW, a Pay 'n Save, a big box store and two Thrifty Drug Stores closer than one mile. I had to make a profit for the store to survive. I did and the owner of the store, Leonard Lasken, was in Los Angeles. He owned maybe twenty All-Med Drug Stores in the San Francisco Bay Area. I saw him twice a year. He would take me out to dinner and a few drinks. The conversation about the store was always brief. Leonard watched only one thing. The profits.

A cautionary word. If you intend on working for a large retail drug store operation, you have to make a profit for them. If they don't make a profit, your job will always be in jeopardy. Like Leonard, that's all they care about.

Okay, I hear you screaming "What about Rite-Aid?" The price for a share of Rite-Aid stock as I write this is $1.03. I don't think the regular retail price of a large Snickers bar at Rite-Aid is that cheap. Rite-Aid is in such a downward spiral that it seems that the company is kept going just so the top executives can continue to get their big pay days. I also believe that Rite-Aid is too big to fail. Can our industry survive the closing of 4700 stores in 31 states? I honestly do not think so.

There are two things that can make earning a profit selling prescriptions very difficult. The bend-over-and-take-an-ass-kicking-from-the PBMs attitude that prevails. Walgreens has refused to execute a ridiculous contract with ExpressScripts. We'll see how that turns out. ExpressScripts has to have contracts with pharmacies where their members live. Otherwise, they will fail and the companies that pay them will have to go elsewhere for a PBM that has a contract with Walgreens. Walgreens is the largest company with more than 8,000 stores. They may have ExpressScripts in a corner.

It is pathetic that Walgreens is leading the way for all pharmacies and bottom of the pit companies like Rite-Aid see the Walgreens situation as an opportunity to get business. I know that the ExpressScript customers will have to go elsewhere if negotiations do not produce a reasonable contract, but it gripes me to see the cheap vultures circling above. If Walgreens wins this one, they are winning for every pharmacy, chain and independent. The dam will be broken. It will be a real battle to watch.

The second aspect that makes it difficult is the drug store company you work for. Inventory control is a joke when the warehouse automatically sends you 12 boxes of diazepam rectal gel when all you sell is two a month. Remember, it is all a cult of personality.

It is really difficult to make and keep friends when you are working at warp speed and do not have enough help. I have been doing this for a long time and I am really talented at the personality game. I know that when a middle-age woman has had to wait too long for her prescriptions that showing a smile and saying, "That blue looks good on you. You look good tonight, Betty" can defuse a difficult moment. The blue may not look that good, but the comment makes Betty feel good. It is not that easy to develop a *cult of personality* when you are telling the people that the waiting time is an hour, but it can be done.

Counseling is difficult also when it is like that, but it must be done. Watch carefully for premium counseling opportunities, when counseling is really important. Otherwise, the 15 second talk on Hydrodocone/APAP 10/660 can be as quick as:

"This is a narcotic. It can impair your ability to drive or work with dangerous equipment." A pause to get her attention. Then you say strongly, "You cannot take more than four of these in a day. 3000 mg a day of Acetaminophen is the maximum. This drug is the number one cause of liver failure and death. This can also constipate you. Eat more fiber."

That takes seventeen seconds. They won't stay with you for longer than thirty seconds, usually anyway.

If it is serious counseling that you know they have to listen to, invite them to the counseling window. Get their attention with these words, "This is serious, listen carefully." our patients respect drugs more than you do. If you have to, tell them that every drug is a poison depending on dosage and frequency. I use this phrase. *If you take enough of this, often enough you are dead so listen to me carefully.*

I have found that more patients than not appreciate being counseled on their prescription drugs. They are not accustomed to it because so many pharmacists are neglectful and don't do it.

Yesterday, I walked over to the register with my bar code. The technician said that she has used this before and has no questions.

I asked, "Has any pharmacist told you to rinse your mouth out after you use this?"

"No, they haven't"

"Well, I'm telling you now."

"Why?"

"To prevent your getting thrush."

"Thank you," she said, "Is this something new?"

"It's the law, Maam.. When I scan my bar code, I am certifying that I talked with you about your prescription."

"No other pharmacists do this."

"I can't speak for anyone other than myself."

"Well, I'll come back when you are here."

There were other patients watching this. They were smiling. If you can develop a *cult of personality*, chain, big box, grocery store or independent, you will be a Super Star. If you do not work for yourself, you efforts will be noticed.

I was Christmas shopping on my lunch break a couple years ago. The non-pharmacist store manager was talking to a man who I recognized as a big wig from Houston. I walked over an introduced myself.

"I believe that you are someone important," I said, "I am Jim Plagakis, one of the pharmacists."

A customer butted in with, "And he is really good too. I always ask him when I need help."

The store manager smiled and said, "He is the best customer service pharmacist I've ever had."

A cult of Personality, my friends

If you remember their names. If you treat them with respect. If you compliment the women. If you girls flirt innocently with older men (be careful with men who still have raging testosterone). If you do all of the things that are natural for you when you like a person and allow a perfunctory relationship, you will be on your way. You will help make a profit for your company.

There is another way to make friends, help people and make money. OTC Counseling. We will cover this in detail in a few pages. Suffice to say, OTC products are real drugs and OTC counseling is real pharmacy.

The Historical Era Continued

If you were a free agent, as I was in a chain situation, you designed prescription pricing. You also typed every single label and every single receipt. You also cashiered and you ordered everything.

The big insurance was Medicaid. If you did not have a clerk assigned to fill out the Medicaid billing forms by hand, you did it yourself. The only PBMs (and they had not come up with the misleading name *Pharmacy Benefits Manager* yet) were Paid Prescriptions and Pharmaceutical Card System (PCS). The clerk filled these out by hand. She calculated the cost from full AWP and added a dispensing fee of $3.60. Can you imagine? This was when the average prescription was around $10.00.

All other insurances (unions mostly) were reimbursed directly to the patient. They paid full retail price and we filled out cards that they sent in to their unions. If you did not have an employee to fill out the cards, you did it yourself.

I emptied the trash. I swept the floor. I made deliveries when it was urgent. I wrote the schedule for all of the employees.

When we filled a prescription, we stamped the front of the paper with a Bates Numbering Machine. When generic selection was allowed by law, we would record the manufacturer. We added the date filled, the filling pharmacist's initials and the price. I always put both the cost and retail price. This was before typing the label and the receipt.

When we refilled a prescription, we had to go to the files and locate the original prescription. We would signify the refill by stamping the date on the back and entering the initials of the refilling pharmacist. We would record if a change from the original quantity was dispensed and also if there was a change in manufacturer. All of this took time.

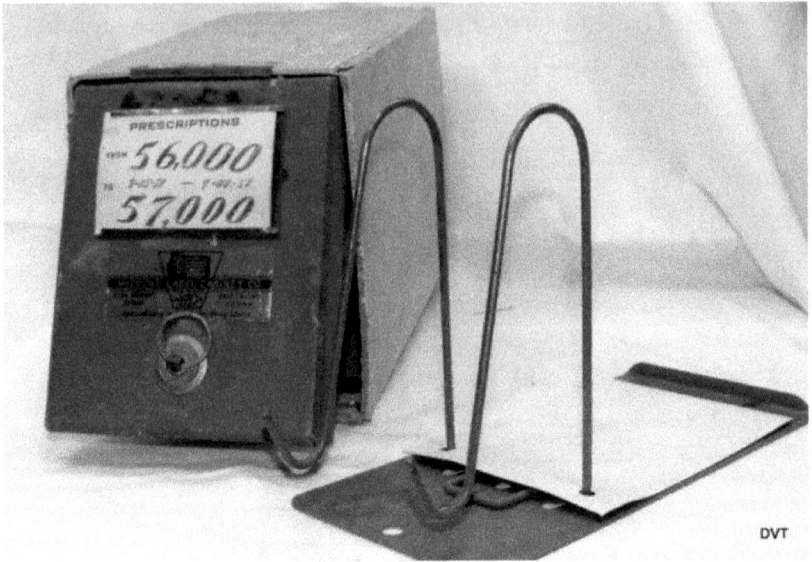

All of that record keeping is now done in a matter of seconds with a new prescription and is accomplished with one click on refills. This is *The Modern Era* of pharmacy. If we had to go back to the methods of *The Historical Era* in the 21st Century, we would be a failed profession, indeed. Remember this when you see *Tech check Tech* coming. Once again, we must adapt if our profession will remain viable and a critical player in medical care. If we can't provide the drugs legally, in an efficient and effective manner, somebody else will. Somebody who is cheaper.

When Pharmacists Were the Captains of the Ship

My first job in a drug store was in 1957. I was 16 years old and I was already thinking that I might want to be a pharmacist. I was a stock boy at Cook Drug Store in Ashtabula, Ohio. There was no question who ran the store. The owner was the boss and he was a pharmacist. The second pharmacist was in charge of the mysterious sanctum, the prescription room. I was in charge of bringing up cases of Rexall products from the basement and filling shelves. The owner, Dolph Hale, taught me the formula to determine a 33.3% mark up. It was easy algebra. I marked the boxes and bottles with a grease pencil.

I didn't know it at the time, but the two years I worked for Dolph were seminal in my education. I learned the basic of the drug store business. I learned that a pharmacist was not above cleaning sinks and toilets. I chipped ice away from the front door and shoveled snow off the sidewalk the morning after a big snowfall. I ran errands to pick up prescription drugs from Wentling's Pharmacy and the other Main Street drug stores (there were six) when we were short of something that we needed to fill a prescription. The drug stores cooperated.

I will always remember when Frank, the second pharmacist, invited me into the prescription room. There was a slight aldehyde aroma. The compounding counter was long and white. There were two balances and a scale for weighing heavier amounts. Frank completed the prescriptions on a counter that faced the front of the store. Two four foot tall multi-colored show globes framed the open window. Dispensing and compounding were the pharmacist's job in those days. Counter-prescribing was the third job.

The prescription room was equally stocked with Pharma manufactured products and bulk drugs, vehicles and chemicals for compounding. Doctors still liked to prescribe custom made products and the pharmacists prepared them. I learned how to make papers and suppositories when I was still in high school. I watch Frank punch capsules and failed when he tried to teach me how to make an emulsion. He told me the simple algebraic formula to get 40% mark up. Every prescription was 40% profit. If we bought it for $1.20, it sold for $2.00. A $6.00 cost would get a retail price of $10.00.

Frank was my hero. He wore a white starched barber-style smock. His shoes were always shined. The music he played on the radio was ethnic Italian. Frank liked opera. My jobs in the prescription room were pedestrian. I cleaned the compounding equipment and mopped the floor. I became intimidated when I dusted the shelves. How could I ever remember all of the drugs? It did not seem possible.

You may not be familiar with the term *counter-prescribing*. Some pharmacists still believe that this is as important in 2011 as it was in 1965. It is unfortunate that younger pharmacists do not think that it is part of the job.

I remind them as often as I can that OTC medicines are real drugs and that OTC counseling is real pharmacy.

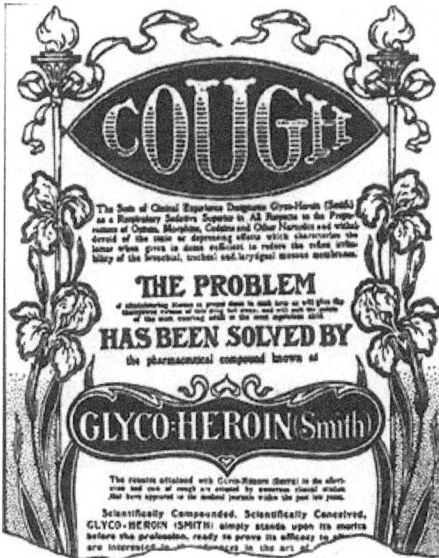

The Golden Days of Counter-Prescribing
Well Before Durham-Humphrey

Counter-prescribing is the prescribing of Over The Counter medicines by the pharmacist. I do not use the word *prescribing* flippantly. People did not consistently get their medical care for minor illnesses like colds, coughs or gastric hyperacidity at the doctor's office. They came to the druggist, the man they called "Doc". Pharmacists made a substantial part of their livings from counter-prescribing.

Counter-prescribing is a carry-over from before the Durham-Humphrey Amendment. There weren't that many drugs in those days. We relied heavily on the official formulas of drugs that were found in the USP and the NF. Pharmacists were still compounders and we were medical practitioners separate from doctors. That is probably hard for you to get. We are so accustomed to following the doctor's lead even when it is not in the best interest of the patient that the idea of a pharmacist being independent is unbelievable.

The Durham-Humphrey Amendment of 1951 changed the game. It formally distinguished between prescription and over-the-counter drugs. Until that time, all drugs could be legally dispensed by pharmacists. No prescription necessary. Hence: Counter-Prescribing. It took more than a decade, but by 1965, my first year as a Registered Pharmacist, we had relinquished our independent role and had securely landed below the nurse in the medical hierarchy. It would take us decades to get back to where we are recognized as an important contributor to the patient's health and we have not gotten back to our rightful spot yet.

Stick with me on this: Some will brand me a heretic for such thinking, but I believe that it is simply intuitive thinking. Modern medical treatment is drug therapy in 90% of the cases. Durham-Humphrey handed our turf over to the medical doctors. We were not clinicians. We were dispensers and they made sure that we stayed in the back seat for over 50 years. The complexitiy of medicine in the 21st Century is changing that. No physician can possibly know more than a handful of drugs. Still, they blow smoke and act as if they were the gods of Olympus.

It is also intuitive that the experts on drugs should be driving this train. The younger doctors seem to see that as the truth. My daughter in law is Doctor of Osteopathy doing her residency in Fargo, North Dakota. A few months ago, she called me for the first time.

"I want a pharmacist with me all the time," she said and explained that she had a pharmacist with her for a month during rounds at the hospital.

"I didn't prescribe one thing, Jim. You guys are amazing. We looked at all of the test results together and I always took the pharmacist's recommendation for drug therapy. Always." Naomi asked me how she could find a pharmacist when she wanted assistance. I told her to call around. There would be a pharmacist who would be thrilled to help her out. I added that it would probably be a younger pharmacist at a neighborhood pharmacy.

"Naomi, the chain store pharmacists are overwhelmed."

Naomi interrupted me. "There are no chain stores here. This is North Dakota."

That's another story.

Senator Hubert Humphrey of Minnesota was the *Humphrey* part of Durham-Humphrey. He was a pharmacist and I was sort of proud of him until I recently examined what his amendment did to pharmacy. It diminished our importance. I suppose that the doctor's lobby, the AMA, supported the amendment. The AMA has been all about protecting their turf for decades. Not many realized until recently that the emperor has no clothes. We now know that doctors are fallible. Pushing pharmacists out to the edge did not benefit the patient.

Frank took me aside one day and explained how modern pharmacy was becoming. He invited me into the prescription room and showed me a bottle of an Abbott product. It was a *Filmtab*. He opened the bottle and told me to smell. It was a pleasant vanilla aroma. Pharma was making medicines palatable. Up until the 1950s, most medicines were liquid and tasted terrible. Frank was very proud of the brand new, ready- made dosage forms.

Frank thought this was a marvelous advance in pharmacy. His starched white barber's smock was more for show now. A few years earlier, he wore it to protect his clothes when he was compounding. I don't believe that many pharmacists could predict the marginalization of our profession after Durham-Humphrey. I cannot imagine that Hubert Humphrey foresaw his profession becoming second-class.

In 1965, regarding prescriptions of legend drugs (Now called Rx Only) my main job as a pharmacist was to dispense and mind my own business. I did what technicians do today. I was not allowed to counsel. That was interfering with the doctor-patient relationship. I used my education when I compounded and when I counter-prescribed. I was a "Registered Man". This was to distinguish me from the men who were legally allowed to do the job of a pharmacist because they had apprenticed to a pharmacist and learned the job on the fly. My first Ohio license in 1957 was an apprentice license. They eliminated that license and entered the 20st Century soon after.

The doctors in Ashtabula, Ohio loved Butisol (Butabarbital Na, McNeil) for women with PMS or menopausal syndrome or other *women's issues*. It seemed like every woman I knew was stricken. I clearly recall believing that it was a drug designed for women. We bought Butisol in four strengths in 1000s. We dispensed it in100 quantities. This was before the BNDD and the DEA. Life had gotten easier for the husbands who wanted their wives to mind the kids, clean, cook, service their manly needs and keep quiet.

A magazine advertisement for Dexedrine
Reason #3, She is bored out of her mind

why is this woman tired?

She may be tired for either of two reasons

Husbands wanted their wives to be sexy, svelte and obedient, so there was a *Stepford Wives* answer in the pharmacy. The doctors were complicit in this subversion. They also believed that women were the weaker sex. It was so bad that a young husband confided to me that his doctor said that his wife should masturbate if his problem with premature ejaculation was not solved. He also suggested to this young man that he frequent the local whore house to *get some lessons*. In was still in my twenties and people confided in me. I have concluded that the only reason I was trusted was because I was a pharmacist.

Of course, the ultimate answer for the man who wanted his wife manipulated like from *The Stepford Wives* was Premarin. We sold thousands every single week.

Of course, women like "Premarin"

The modern image of the pharmacist as a dispenser of a product is not accurate. Doctors don't know that. Nurses don't respect us. Most patients believe that all we do is fill prescriptions. You have heard this many times: *Why does it take so long to take pills from a big bottle and put them in a little bottle?*

We do not have any modern organization that promotes and enhances our image. There is no organization that portrays pharmacists as important purveyors of medical care. Why not the APhA?

Where is William S. Apple when we really need him? He was the APhA president when I was a member and when I had faith in our national association. Apple promoted a third class of non Rx drugs that would have to be kept behind the counter and sold only by a pharmacist. This was in the 1960s. Of course, the doctors saw this as their turf and fought it. If you have been paying attention, the doctor lobby fights us every time we want to expand our responsibilities. It doesn't matter if it will benefit the patient. If the doctors see any diminishing of their power, they will fight.

William S. Apple created a legacy that has been lost. He was a brilliant leader. He was a tireless advocate of pharmacy and pharmacists. There has been no one close to him. I trusted that he would act in a manner that would benefit me personally. I haven't felt that way about APhA for a very long time. I doubt that many pharmacists do.

There were only a few chain drug stores when I took that stock boy job. Across Main Street was Standard

Drug. Further down, one block to the north, was Marshall Drug. Both were managed by pharmacists. In 1964, my college roommate took a job with Walgreens in Chicago. He started out as the Assistant Manager. He was a pharmacist. The Manager was also a pharmacist.

In 1966, I worked for awhile for SupeRx which was a Kroger company. I was the Assistant Manager and the Manager was a pharmacist. That is what pharmacists did in those days. We managed drug stores. We were pharmacists and we were drug store merchants. We filled prescriptions and we also managed the cosmeticians. We didn't run the lunch counter, but we managed the wait staff. We wore a business hat as well as a professional hat.

I can clearly recall my first job as a Registered Pharmacist. I was 24 years old and wet behind the ears, but the owner of the store gave me the title of Assistant Manager. With the title came responsibilities that I was ill prepared for. I enjoyed the position, but when I saw two employees sitting at the lunch counter for an extended time, I was much too heavy handed when I told them to get back to work. People who liked me at the owner's Saturday night house parties seemed to hate me at the store. I finally said to the owner, 'I can't do this. I just want to be a pharmacist. I am a bad manager."

He said, "You are not a bad manager. You are young and inexperienced. It is not as serious as you make it. You will learn."

I found out one very interesting young man's benefit of being a pharmacist/manager. The woman who ran the large photography department was thirty-something and stunning. She was a sex object. She dressed like one and she acted like an object. This was only a few years after Sandra Day O'Connor got out of Stanford Law School and was offered a job as a secretary. In 1981, O'Connor became the first woman in the 191 year history of the United States Supreme Court to become an associate Justice.

This woman from the photography department was a country girl. She had a high school education. There were no good jobs for uneducated women. Her sexuality was her stock in trade. Everything about her commented on her sex appeal. I loved her husky voice and how she made it sing in a way that made me think that everything I said was clever. Her hair was always in place. Her makeup was planned to accent her blue eyes. I can remember the aroma of her fragrance. A feminine Lily of the Valley perfume that tasted salty when I kissed her neck. Her body was fit, but rounded and soft enough in the right places.

She and I got along very well and when I invited her to go out after work to a popular bar for a sandwich and some drinks, she agreed. We had a terrific time. I drank Scotch on the rocks and she was drinking Manhattans. After, we stopped at a place for some coffee and we gabbed. I was too innocent to pick up on what touching me and letting her fingers linger on my arm meant. When we got to her apartment, she invited me in.

I had been through a very long dry spell and was a horny 24 year old. She popped a couple beers and excused herself. I let my imagination run. I quickly planned my moves, but I was not prepared for what happened next. I can still remember the Look magazine on her coffee table. Jacqueline and little John John Kennedy were on the cover. After a few minutes, she came out from her bedroom wearing a smile and a negligee. She was naked underneath. My mouth became immediately parched.

I knew that a musician who spent most of his time in Hollywood called her his girl. He was the band leader at Desilu Studios. When he had a break from his schedule on the coast, he would take a train and return to Ashtabula. He would stay at his mother's and spent most of his days sitting at the end of the lunch counter at Wentling's Drug Store, a few feet from the photography department. He would smoke cigarettes, drink coffee and watch her. I asked about him.

"He's not here, but you are," was her answer as she pressed against me. She kissed my neck and that was when I showed just how inexperienced I was. I returned her kiss, but was mortified that my manhood was quickly wilting. I did not know what to do with a woman who was sexually aggressive. Woman were supposed to play coy and hard to get.

I stepped back a step and asked, "I am sorry, I. I..why…Why me?"

"You are a pharmacist," she said, surprised at my question. "I don't do it with just anybody, Jim. You're somebody."

I failed the test. She did not give me a chance to seduce her. She ruined my plans and I couldn't do it. The next day, I bought her a cup of coffee and we smoked and talked for awhile. I confessed that I had been intimidated by her manner and told her that maybe we could have drinks again some night. I will never forget her answer.

"Any time you want," she squeezed my forearm. "Just let me know. I'll help you all I can."

I never did follow up. I got my California pharmacist license a few weeks later and was distracted. There was a coldness between us after that. I think now that she believed that I had rejected her. I had learned a hard lesson. If you are a boy, be very careful around a woman.

I am relating this story because I did not realize how I was perceived until that night. I knew that I deserved some deference in the professional world because I was a Registered Man, but I did not understand how my standing filtered out into the real world. I watched people carefully after that and, over the decades, it is very clear that pharmacists are looked at with esteem. Some of this was a carryover from the days when pharmacists counter-prescribed. I clearly remember being told often that my position was respected.

"I trust you more than I trust the doctor." I have heard that often in my career. I believe that much of that sentiment is because I am free. Of course, they would want the person giving free advice to be competent. Some people actually believed that my skills were superior to the doctor's. In the end, I do not diagnose, but I do know a lot about drugs.

The pharmacy owner was right. I did learn how to manage. I was more even handed in my first job as a store manager in California. I was in the job for less than a month when I discovered that a long time employee had falsified her time card. I waited until the next pay period. She did it again. I had learned that being a bulldozer was not the right tactic. I was proactive and called the union representative. I showed him the time cards and told him that I did not believe that two time cards in a row was an accident. I explained that I was going to fire her for cause and expected that the union would not defend her. She was a thief.

The Dark Era
An Extreme Example

Non-pharmacist MBA Masters of the Universe discovered that they could use the pharmacy computer systems to spy on the pharmacists. CVS has made this into an art form.

CVS pharmacists said nothing. They did not complain or rebel. Like many of the conditions pharmacists work under, they waitied an interminably long time to start bitch and complaining. Some companies call the timers *metrics*. I have been told by CVS pharmacists that the company is getting downright medieval about the *metrics.* They write pharmacists up for too long wait times. I believe that non-pharmacist managers and pharmacy district managers have been told to use these words.

If you can't do the job, we'll find someone else who **can** *do the job*

CVS pharmacists, with gallows humor, like to morph it into this:

If you can't do the job, we'll find someone else who **can't** *do the job*

The most troubling phenomenon of the *Dark Era of Pharmacy* is the power that we have given to non-pharmacists. There are non-pharmacist executives making decisions that affect all of us profoundly. They never ask our opinions. They are enamored with non-pharmacist MBA Masters of the Universe who crunch numbers and make a business advantage the basis for all of their programs and decisions. They are often very heavy-handed and seem to consider their professional staff to be no better than piece workers on a factory floor.

The example of Kelly Hoots, a Pharmacy Manager in North Carolina, comes to mind. Kelly was fired by CVS because he closed the drive through for cause. When the non-pharmacist manager had a hissy fit and opened the drive through, Kelly closed the pharmacy at midday. A week later, he was fired.

I was incensed by this arbitrary punitive action by CVS. Kelly is a young man. He and his wife have triplet toddlers. They were sick with worry. I wrote this letter to the North Carolina Board of Pharmacy. The Secretary wrote me and asked that I refrain from writing board members. A formal complaint against CVS had been filed. In order to prevent a member having to recuse himself/herself, it would be best to be patient.

July 11, 2011

North Carolina Board of Pharmacy
6015 Farrington Road, #201
Chapel Hill, NC 27517-8154

This letter is a request that the board review the circumstances of the firing of Kelly Hoots (North Carolina licensed pharmacist) by CVS. We request that the board give consideration to the facts and intercede on Mister Hoots' behalf. Mister Hoots was a trusted and competent pharmacist in charge for CVS. When he made a decision in an effort to assure patient safety, Mister Hoots' authority to close the drive-through was illegally usurped by a non-pharmacist. At this point, Mister Hoots realized that the situation in the pharmacy was dangerous and untenable and he asserted his responsibility to make sure that patient safety came first and he closed the pharmacy. The board will recognize that this was his legal responsibility. He had no other viable choice.

Mister Hoots reports that at hour seven of a fourteen hour shift, he had already reviewed and verified over 300 prescriptions. His intention was to counsel appropriately where the law required.

That is one prescription every 84 seconds, with inadequate help. There was no uninterrupted break for nourishment or even a bathroom break. Clearly, Kelly Hoots was compromised. It is to his credit that he took action as North Carolina Pharmacy Law requires before a patient was harmed.

The Board of Pharmacy is mandated to regulate the practice of pharmacy in a manner that protects the public from harm or potential harm. CVS consistently operates in a manner that not only disregards North Carolina Pharmacy Law, they flaunt the law.

I specifically point to: -21 NCAC 46.1804(b) and -21 NCAC 46.141(b). The non-pharmacist store manager has absolutely no authority within the pharmacy. Mister Hoots determined that the conditions obviated his providing safe and effective delivery of prescriptions and made complying with counseling legal requirements impossible.

CVS is well known as a Goliath company that bends and breaks the rules. It seems that this is the corporate culture. The Federal pseudoephedrine sale recording requirement is a case in point. CVS agreed to pay a $77 million fine for neglecting to abide by the law. CVS flaunts the law at every turn. The executives in Rhode Island seem to be invested in only one thing, the bottom line and they will do anything, legal or illegal to get what they want.

It is a sad commentary that such a big player in the retail pharmacy universe gains a competitive advantage by cheating.

In the case of Kelly Hoots, CVS broke the law. It cannot be any more clear. They cheated and they are probably cheating all over North Carolina. In this one case, they must be held accountable.

I urge the North Carolina State Board of Pharmacy to waste no time in taking effective measures that will give Kelly Hoots and his wife some relief as well as assuring, by example, the citizens of the State of North Carolina that disregarding the law, by any drug store company, large or small, will be punished.

Jim Plagakis

The Veneer

On page 6, I began a discussion of the non-professional duties that take up most of our time. For our purposes, we will call this a coating or *veneer* that is a heavy, ponderous covering over of the practice of pharmacy. I like to label it "The Prescription Mill". Included in "The Mill" are the filling of prescriptions, the timers (metrics), reports, drive-through lanes, cash registers and any other good idea that the non-pharmacist MBA Masters of the Universe at the big companies can up with. None of these aspects of "The Prescription Mill" have anything to do with the practice of pharmacy. Or, said differently, the practice of pharmacy is being covered up artificially by a layer or veneer that is called "The Prescription Mill".

There are no laws governing the managing of "The Prescription Mill". There are plenty of laws dictating the practice of pharmacy. You are not going to get in trouble with the state where you practice if you don't get prescriptions out in a certain number of minutes. The state board is not concerned with how fast you make it to the drive-through. Can't get an insurance claim to adjudicate? The state board could not care less. The board is in existence to protect the public from danger that emanates from poor pharmacy practice. I will argue that the *veneer* can and does interfere with the practice of pharmacy. Running "The Prescription Mill" consumes the time of the average retail pharmacist to the point of distraction. Because of the Mill and because the Mill is the only thing that some companies care about, the pharmacist neglects his real professional responsibilities.

There is no legal publication that deals with "The Prescription Mill". There are thick books of pharmacy law. If you are a licensed pharmacist, you know the law. At least, you better know pharmacy law. The boards are not kidding when they find out that a non-pharmacist let himself into the pharmacy when no licensed pharmacist is present.

This happened in Washington State. It was related to me by a state board investigator whom I consider a friend. The non-pharmacist store manager was unhappy with the pharmacist in charge. He had built a case against the pharmacist. In order to compile his evidence, he had to enter the pharmacy when it was closed. His accusations were ambiguous, but within the hierarchy of this very large grocery store company his complaints would have legs.

The state board investigators in Washington State have police authority. When she got to the pharmacy around midnight, the non-pharmacist store manager was already inside the pharmacy. He had opened cabinets and drawers to make the investigation easier, he thought. This was not a drug issue. It was a matter of the non-pharmacist manager building a case to get rid of the pharmacist in charge. If I remember correctly, the PIC was guilty of something minor, but when the store manager broke the law, his charges became moot.

The investigator asked him, "Do you come into the pharmacy often when it is closed?" She told me that she kept a smile on her face.

"Only when it is important." He started to outline his complaints against the PIC. He had paperwork on the counter. She told me that he was wearing a work out costume and was smoking a cigarette. She also commented that he was wearing a toupee, a bad toupee. He was a short man with a pot belly.

"Where did you get the key?" The investigator is a pharmacist and she told me that she was so pissed off that there was smoke pouring out of her ears. She hid her anger. Her appearance and her demeanor were entirely professional. She had come prepared and had put a voice recorder on the counter. It was running.

He looked surprised at the question. "It is an extra key. I keep it locked up. No one has access to it." Oh, what a good boy you are.

"Do you keep it in the safe?" She was being careful to ask the right questions.

"No one else has access to it." he repeated. "I keep it in my private lock box."

"How many times have you used the key to get into the pharmacy?"

The manager was thoughtful. He counted off on his fingers. He took a breath. "Maybe a dozen times. I like to keep on top of all of my departments. I inspect the pharmacy just like I inspect the butcher shop."

"For the record, tell me your name and your position with the company."

He did as she asked and then asked her why she had requested the information when she already knew the answers.

The investigator indicated the voice recorder and then calmly explained that his entry into the pharmacy was illegal. She announced that she was going to write a citation on his behavior and that she was also citing the company. There would be hefty fines. She added that she was going to request the board of pharmacy lawyers to explore further sanctions.

The non-pharmacist store manager got all puffed up like only a community college attendee could get. "My company's lawyers won't stand for that. Are you going to investigate my complaint about the pharmacist now?"

Then she explained that the pharmacy manager was the authority in the pharmacy. "You have no weight in the pharmacy, Sir. The law gives the pharmacist on duty all of the power. You may be the store manager, but you are not a pharmacist."

"But he…"

"…I read your complaint, but you have no standing in the matter. You admit that you have been entering the pharmacy illegally. You could have very easily arranged evidence to make the pharmacist look bad."

"So, what are you telling me?"

"To the point, you, Sir, are busted."

Pharmacy jurisprudence does not even address the veneer of artificiality that covers the practice of pharmacy. The boards of pharmacy are mandated to regulate pharmacy practice so civilians are not harmed. The boards don't care that you don't get to eat lunch. They don't care that you don't have enough help. They don't care that you don't have a cashier after 7:00 PM. They don't care that you work fourteen hours straight, with no rest periods.

If the conditions you work under endanger the public, the boards must care. The problem is that the boards are like the three monkeys. *See No Evil, Hear No Evil, Speak No Evil.* How else could they miss it? They would act immediately if a pharmacist had a couple martinis at lunch or smoked up before work. If you got caught taking a lorazepam during the evening rush, you could be cited, possibly lose your license. What about the pharmacist who is filling a prescription every 80 seconds and is in the 12th hour of a 14 hour shift?

You are a medical professional and your most important obligation to the patient is the *duty to warn.*

One of these days soon, a pharmacist is going to be cited for not complying with the state and federal laws requiring counseling. The pharmacist will be cited and there will be a license suspension and a fine. The drug store company will be cited and fined. The pharmacist will be required to complete the MPJE. The company's Pharmacy District Manager will be required to complete the MPJE. The store's pharmacy license will be suspended for a period of time.

This is serious business. If you do not pay attention to pharmacy law you are practicing under a cloud. Only an idiot would flaunt the law. It happens thousands of times a day. I work part time in a state where the board requires pharmacists to certify that they have counseled on all new prescriptions. In my company, each pharmacist has his or her personal bar code. At the register, the process stops when a new prescription is in the sale. It does not proceed until a pharmacist's bar code is scanned. One/ enter certifies that counseling was accepted. Two/enter says that it was declined.

I run back and forth between the pharmacy counter and the register all day long. The other pharmacists who work in the store regularly print off their bar code and leave it by the register. The technicians do the scanning and the pharmacist is usually not even called.

What gets me is that the pharmacists do not even seem to know that they are gambling and the stakes are their careers. If they do not warn and the patient is harmed, they could be held liable.

No judge or jury will give the pharmacist a free pass. If a woman in her middle years has a stroke because the pharmacist failed to counsel, a judge is likely to give instructions to the jury that are downright medieval. There would be no passing GO. There would be no Get-Out-Of –Jail-Free card.

I worked with a pharmacist is California who was found to be negligent on an insulin prescription. The child was given what was on the label. The doctor had made an error. It was ten times the prescribed dose. The child went into insulin shock and had to be hospitalized. The parents sued both the doctor and the pharmacist. The doctor had malpractice insurance. The pharmacist did not.

This pharmacist was a single man. When he was unable to pay the judgment, he went bankrupt. This was in the late 1970s, before OBRA 90 and state counseling laws. Can you imagine the judgment if the child had been permanently harmed and the pharmacist had not counseled when mandatory counseling is the law? What if the child died? Another Eric Cropp? Is it logical to assume that during the counseling process the high dose may have been questioned?

There would be pharmacists who are such weasels that they would try to blame the technician. She engages in fraud when she scans the bar code. It reads below the pharmacist's name:

This barcode is intended for use ONLY by the pharmacist identified above. Use of this barcode by any other person other than the pharmacist

identified above will result in disciplinary action,
up to and including termination of employment.

That is a straightforward warning. It is not ambiguous in any way. If a patient is harmed by their medication and they were not counseled, a can of worms is opened and a good career will be in jeopardy. The technician will plead that the pharmacist directed her to commit fraud, but she will still lose her job.

A jury will crucify you and the judge will look smugly on. You are the professional. That designation conveys responsibility. It is irresponsible to disregard any law of the profession.

You are guilty and the jury awards the woman a lot more money than your insurance will pay. Don't look at your company. They will say that the company expects all pharmacists to obey both the spirit and the letter of all pharmacy laws. The company attorney will show the jury and the judge Exhibits A thru M. They are all statements that you signed promising to obey the law. What about this is so hard to understand? I certainly hope that your throwing of the dice doesn't get you into trouble. My hope, by the way, gets you nothing.

Stupid Moves by Non-Pharmacist Executives

It does not matter what the choices, programs or decisions made, if they regard pharmacy, pharmacists must be in the mix.

I cannot imagine pharmacists coming up with the 15 minute guarantee, the metrics or allowing non-pharmacists to manage the professional staff. But, they do. Imagine this about any of their bright ideas.

*I*magine the discussion when the executives and MBA types sat at a meeting at a large grocery store chain. These decision-makers had not been brought up in the pharmacy culture and honestly did not respect pharmacy as a profession. As far as they were concerned, they were dealing with a commodity and all they wanted to do was to sell more commodities.

They saw themselves as grocery store versions of Tom Wolfe's *Masters of the Universe.* They were going to squash the competition and they were going to utilize the professional department and the professional staff to do it. In the end, they were all interested in only one result and that was their yearly bonus.

Who cares that pharmacists are highly educated medical professionals? What does it matter that pharmacy has been losing respect among its practitioners over the last two or three decades? Why is it important that the public perceive that an important professional service is provided when prescriptions are dispensed? We have been competing on price since 1972. Why change now?

Listen in to the meeting in the executive suite. There are croissants and coffee on the sideboard. The CEO liked little sandwiches made with cream cheese and cucumber. A young woman, wearing a little black business outfit, was on her way up in the company. She knew from experience that her sexuality gave her a leg up with the CEO.

The MBAs had all eaten heartily before the meeting. They were thirty-somethings and the cucumber sandwiches were the subject of jokes when the boss wasn't around. The appetites and preferences of the CEO, a smallish man from Rhode Island, were always deferred to. His mother always told him that a light lunch made for an easy afternoon. He never questioned his mother. He rarely questioned anything, actually.

"The pharmacists won't say a thing," said an accountant who had gotten his Masters degree at a night school Executive MBA program. "They never complain about anything." He was in nicotine withdrawal, but knew that he daren't light up a Sherman's until the CEO took out a big cigar. He fidgeted nervously with a paper clip.

"Have they ever complained about anything?" chuckled the Vice President of retail operations. Then his smile hardened. He looked right at the Vice President for Pharmacy Operation, the only pharmacist at the table. "Pharmacists are just overpaid clerks."

"Oh, I've gotten complaints," said the Vice President for Pharmacy Operations. "But they are scattered. Ironically, it is the best pharmacists in the company who have problems with some of our programs." He looked at the CEO. 'If our best pharmacists rock the boat, we would..."

"....They aren't going to rock anything," interrupted the CEO. He was tired of talking about the pharmacists. They were no more important than the bakers. "Do we have to pay them that kind of money?"

The VP of Pharmacy sighed. This was a question at every meeting. "If we want to have pharmacies we have to have pharmacists and we don't pay any more than any of the big players."

"All pharmacists care about is that big paycheck," argued the young woman. "They won't say anything about this."

Avery, the only minority at the table laughed. "They won't quit and they are afraid of being fired. They will toe the mark as always because they have debts."

"Big debts," added a Division Manager. "I talked with a kid last week. He isn't thirty yet and he has a house on the lake. His kids go to a private school and his wife drives a brand new Lexus.expectantly"

"I like this free diabetes drugs idea," said the CEO. He was nibbling on a cucumber and cream cheese sandwich. "I like it a lot. The generic drugs you are talking about are dirt cheap. Go ahead with this. Give them away." He was looking right at the VP of pharmacy. "We'll get more good will for the buck than we would ever get with advertising that costs a hell of a lot more." He smiled at the young woman sitting beside him. He liked her. She made him feel young. "We can't forget that we are in the grocery business."

"The pharmacists will get onboard. Eventually, Wal-Mart spun the $4.00 prescriptions as a contribution to public health."

"We can do the same, but bigger than they did and right from the start. Let's put some pharmacists in our newspaper ads. Free prescriptions, it can read."

"Pharmacists are still professionals," the pharmacist said. "We have to show some respect. They don't like their image being cheapened."

"We do not care what the pharmacists want or do not want," said the Retail Operations Vice President. "I suppose you care because you are one of them."

"Yes, I am a pharmacist." He wondered how long they were going to include him in these meetings. The pharmacy was never going to be anything more than a tiny department in the corner that cost an enormous amount of money to run. His last position was with a big drug store chain that considered the pharmacy its cornerstone of its business. He missed those days.

His kids were grown and out of the house. His wife had been complaining about his hours. Maybe it was time to cut and run. He had bought the little house in the beach town in North Carolina for retirement. The owner of the small town drug store has offered a part time job. He hadn't practiced pharmacy for over twenty years. He wondered what it would be like, back behind the counter again. People respecting him instead of being treated like a piece of crap by these grocery store Bozos.

"This is a good plan," said the CEO. He had taken out his big cigar. This was the signal that the meeting was over. The matter was settled. Everyone gathered their papers and stood to leave. Except the pharmacist. There was a melancholy smile on his lips. The group as a whole looked at him expectantly.

The pharmacist picked up a little cucumber and cream cheese sandwich and looked at it. "This is a sad excuse for a sandwich, Tommy." He smiled at the CEO. "Whoever told you that this was good food?"

"What are you doing, Mister Johns?"

"This is my way of telling you that I think that you and everyone else standing there are going to fail. The pharmacies are the class of the company. You cheapen pharmacy at every turn and I am not going to be part of it any more. I am going to cash in my stock options, take my golden parachute and have a good life."

"You don't have a golden severance package," the CEO spat. "We don't give golden packages to ...a...any pharmacist."

The pharmacist smiled. "You should read contracts before you sign them."

If only, just once, that could happen.

My first job with a non-pharmacist store manager was at Thrifty Drug Store in Pleasant Hill, California.

I had never seen a drug store this large. It was close to the size of half a football field. It carried everything from car batteries to plastic furniture. I was only a few weeks removed from North East Ohio. I had arrived in a culture that valued everything cheap and disposable. Thrifty carried shoes and lamps, casual clothes and a huge cosmetic department. The Liquor Department was as larger than any free-standing state liquor store in Ohio. The fishing supplies ran down an entire aisle, on both sides. The ice cream cone stand was much more popular than the pharmacy. The pharmacy was located at the back of the store, right in the center. This was before pharmacies had gates that could be closed. That meant that the pharmacy had to be open every hour that the store was open and that the store could not be opened without a pharmacist on duty. This gave pharmacists a lot of clout with the company.

If the pharmacist does not show up, the store could not open. I can remember coming in late only once. Customers were waiting outside the door. The store manager was fuming. He made threats, pointing his finger at me.

Angelo Lazzareschi was the Pharmacist In Charge. Thrifty did not have Pharmacy Managers. Lazz told me not to worry about the store manager. He reminded me that I belonged to the union. I learned quickly that I was protected from retribution from the manager over just about anything that happened.

I had my first taste of a young man's game. The fourteen hour shift. I worked at least one long day a week. Pharmacists were hourly employees. Union rules meant that they could not schedule us for less than eight hour shifts. All hours over eight hours in a day or forty hours in a week were paid at time and a half.

The first time the Hells Angels came into the store, I felt like I was in The Twilight Zone. The Angels were mythical bad boys who were thousands of miles from my hiding place in Ohio. Then one day, they were right there, four of them, staring at me. Scruffy hair, beards, ear piercings, leather chaps, sleeveless denim jackets and muscular upper arms. Then, one of them pointed at me and indicated that he wanted me to come to the counter.

I asked how I could help him. I suppose that I had seen "The Wild Bunch" too many times and had read too many sensational stories about the Hells Angels and was expecting something very bad to happen. There was a waist high door with a lock on the inner side. It would be no problem to reach over, unlock the door and invade the pharmacy.

The Hells Angel in front of the group handed me two prescriptions. He had what I interpreted as a mean, aggressive look in his eyes. I had heard that the Hells Angels were usually high on drugs and liable to act unpredictably. I stepped back.

"Take the fuckin' prescriptions." I recall those words clearly. I took the prescriptions carefully and looked at them. One was for Pentids 400 (Penicillin G 400 units/5 ml Squibb) and the other for Phenergan Expectorant with Dextromethorphan.

I asked, "Do you want me to fill them?"

"Is there another reason why I would give them to you?" All of a sudden, this man who I had considered to be dangerous was just another concerned father. He looked at his comrades. "Can you believe this guy?"

There was a lot to be concerned with. Bacterial bronchial infections were common in the 1960s. Children and adults were coughing and spitting all over the place. We dispensed Penicillin G suspensions, Tetracycline syrup, Gantrisin liquid and cough syrups by the dozens.

"Is that medicine any good?" One of the other three Hells Angels asked. He eyed me cautiously.

"As good as any," I said. "It's for a bacterial infection."

"If it doesn't work, we'll be back to see you." A heavy-set blonde guy added.

"Why would you blame me?" I asked. "I'm the pharmacist, not the doctor. I just prepare it, I don't prescribe it."

"So you're telling us to get the fuckin' doctor if my baby doesn't get better?"

"Yes," I agreed, "Get the doctor." This was a first for me. I was used to protecting the doctor. I asked myself why I did this. I decided that if anyone was going to take a roughing up if the Angel's baby did not get well, it would not be me.

Apparently, the Pentids did the job. There were no news reports of this doctor's badly beaten body being found in an irrigation ditch in the Sacramento River delta.

The Hells Angels were actually a revelation. They were much bigger people than the popular misconception of them being one dimensional trouble makers. They were also fathers of little children and good friends. In many ways, they were as normal as I was. Their old ladies may not have been their legal wives, but they were monogamous. That made their women more normal and faithful to their men than my wife.

I thought I was ready for anything after the Hells Angels until I spotted an older Roman Catholic Priest walking up the aisle wearing flip flops. He was about six foot five and was accompanied by two nuns wearing knee-length skirts. They came sauntering down the aisle, stopping to look at the laxative offerings. This was, by the way, the first store I worked in that was self-service. Wentling's Pharmacy had most of its stock behind glass display cases, in wall cabinets that had sliding glass doors.

I had never seen Roman Catholic nuns wearing anything other than full habits, with skirts that covered the ankles. These nuns wore black pumps instead of the heavy, man-style shoes that I had seen nuns wear in the past.

When they got to the pharmacy counter, I took a good look at the priest. He had a golden tan and a two day beard. His dark grey suit was cool linen. He handed me two prescriptions. They were for the nuns. A product called Norodin. It was methamphetamine Hydrochloride, Endo. Apparently, the nuns were depressed. They certainly did not need to lose weight.

But, perhaps, these were not their first prescriptions for this product. I went to Bing Images and look what I found. Nuns, ladies out of the dark?

In those days, pre BNDD & DEA, getting a prescription for amphetmaines was as easy as asking the doctor. Women became addicted and didn't know what was happening to them when they tried to quit the drug. To make matters worse, the doctor and pharmacist had no idea also.

I defer to my better judgment and will not make comments on the mystic and mystery of the Roman Catholic Covent on Ygnacio Valley Road. I will relate the popular story of Clark Gable and Carole Lombard, instead.

I lived on Ygnacio Valley Road in Walnut Creek. In 1966, there was nothing on the road other than a small medical center, two strip malls and the apartment complex where I lived. Across the street, up the hill toward town was a large stucco house with a tall arched brass front door. In 1942, it was owned by Clark Gable and Carole Lombard. I searched for images and this was the only picture of the Gable / Lombard house Google gave me.

Gable was here when the TWA Western Airlines plane crashed into the mountain near Las Vegas killing Carole Lombard and everyone else aboard. Gable left the home that night to go to the mountain in hopes that there were survivors. It was not to be. He never returned to the house on Ygnacio Valley Road.

When I lived near this gorgeous house, it was a convent. It was where the two nuns lived. I was never able to figure out if the priest lived there also. I offered to deliver some prescriptions to the house one evening because I lived just down the street. A Hispanic woman answered the door. She was a servant apparently. I told her that I wanted to deliver the package to the recipient, but she was much too accomplished. She closed the door in my face after giving me the exact price and taking the package.

That day at the pharmacy, the priest handed me the two prescriptions for the nuns and I said, "Hello, father."

He gave me a look and asked, "Are you Catholic, my son?"

I told him that I was not a Catholic, that I had been raised in a Protestant faith. "My family's church was the Congregational Church."

"Harvard," he sneered, "Hardly a religion at all." He studied me. "Why did you call me Father?"

"Well, I took a summer Trigonometry class at Gannon College in Erie, Pennsylvania." I paused, "Gannon is a Catholic school."

He nodded, "I know Gannon," he said.

"My professor was a priest. I called him Father."

"Why did you call him Father if you are a Protestant?"

"Because everybody else called him Father."

The priest studied me for a long time. "You don't have to call me Father. If you aren't sincere, I would rather you did not call me Father."

Both nuns smiled broadly. He caught them and they immediately frowned.

"What should I call you? Mister?"

He laughed. "Okay, call me Father."

The nuns got prescriptions for Norodin every month until I left the store and moved on. At the time, I didn't think much of it. There were no Controlled Substances. I dispensed Dexedrine, Dexamyl, Preludin, Biphetamine and Eskatrol every day, all day long. Then, to put them to sleep, they needed sleeping pills. Uppers in the morning and downers at night.

Marilyn Monroe, all by herself, gave barbiturates like Seconal and Nembutal a very bad reputation. Pharma came up with new, *harmless* sleeping agents. Doriden and Quaalude were the leaders.

Have you noticed how women were perceived forty years ago? Helpless, hopeless and dependent. I actually believed that women were fragile. My wife was a ball-buster and I was still gentle with her, even after it was clear to me that she had made me a cuckhold.

I am sure that you have heard of "Ludes". If you haven't, take a Google tour for Quaalude. It wasn't that long ago that pharmacists were simply legal drug pushers.

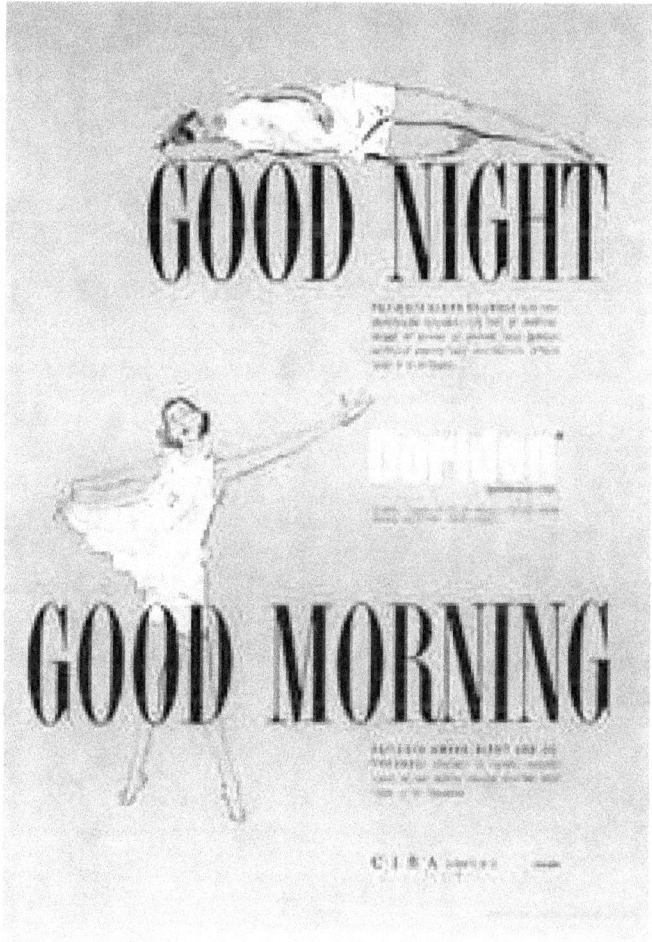

This was Doriden, only one step removed from Quaalude.

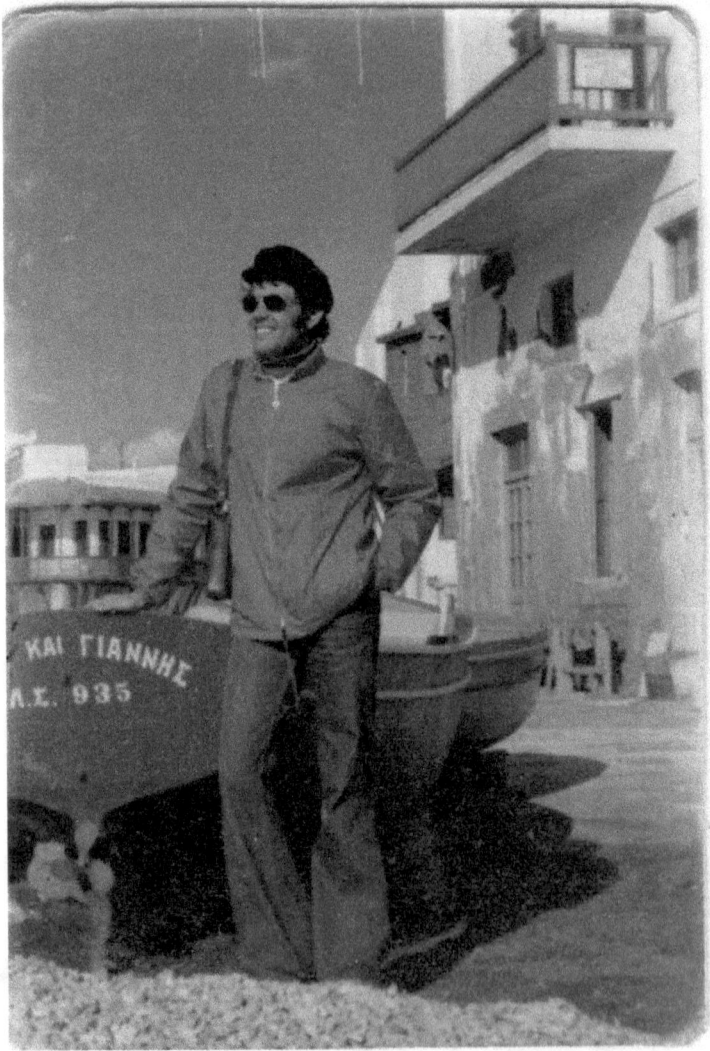

The Island of Mykonos, Greece 1976

History Tour

I have been giving you this pharmacy history tour from my viewpoint because I believe that it will be helpful for all of us to recognize how we got to where we are. I believe that the modern era of pharmacy began when pharmacy work became computerized. We entered pharmacy's epoch of darkness when non-pharmacist manager types discovered that they could spy on pharmacists with productivity programs. We will get to that. I will talk in depth about all of the darkness that is prevalent in the job of working in a retail pharmacy. First, I want to take you through what were the best years of my career.

First, some good advice. Stay out of debt and if you are in debt, do everything and anything you can do to get out of debt.

This is me on the island of Mykonos in the Aegean Sea in December of 1976. I had quit my job working as a store manager/pharmacist for All-Med International after seven good years. My first marriage had been a disaster. For a year, I drove to South Shore Lake Tahoe and played Blackjack every single weekend. I won a lot of money, paid off all of my first wife's bills and had a substantial wad of cash. So much that I did not have to take a full time job for over five years.

Mykonos was a brilliant experience. I was so far away from working in a pharmacy that I felt like I was a different person.

I am going to give you some advice. If you are young and single and have avoided debt, take some of the money you have saved and go live some life. Quit and go live. You are a pharmacist for goodness sake. You will find a job when you come back.

I am acquainted with a young couple from Sandy, Utah. Colin and Sarah are married. They are both pharmacists. For years, they lived their lives to the fullest. They would work six months and then travel in Europe for six months. They always found jobs when they returned. I asked Colin about their peripatetic lifestyle a year ago. He told me that he and Sarah were staying with jobs longer because the pharmacist shortage wasn't as acute in Utah as it was. I did notice, however, that he and Sarah participated in a medical mission to Haiti, I believe.

I believe, learned from experience, that there is nothing that will sustain your spirit better through long years behind the pharmacy counter than a brief love affair in an exotic place. For me it was a young Irish girl who was looking for an experience impossible in Dublin.

I do not believe that you will find a five dollar a night hotel room with a shower on Mykonos in 2012. I visited in the winter, the off season, and I was among only a handful of visitors. I have been told that there is no off-season on Mykonos anymore.

I would go with another American down to the dock every day to welcome the daily ship from Piraeus and to see if there were any Americans onboard. We would sit in the sun, drinking a beer or a coffee and smoking acrid Greek cigarettes watching the boat come in. It was too large to enter the harbor, so they offloaded the passengers onto smaller boats that brought them to the quay.

There rarely were any Americans. I suppose that that was why I eventually decided to return to the United States. That and because the Irish girl had decided to return to her parents' home in Dublin. She gave me a last kiss and said that it had been brilliant, but she had a serious need for confession and absolution.

The Unbearable Lightness of Being is a novel written by Milan Kundera. It is a thoughtful, sometimes painful, but always brilliant story. The movie was very good, but the book could not be matched. I found myself on Mykonos living this unbearable lightness. I was not entangled with anyone, anywhere. I did not have to be anywhere at any given time. Nobody cared that I was spending my days walking around a white-washed village on a Greek island. The island had 365 tiny chapels, one for the Saint of every day in the year. Every day, I walked to the chapel of the day and watched the visitors come and recite their brief prayers. My mother would have been upset if she never heard from me again, but no one else. Well, probably my dad and brother. I felt that detached and that light.

I had spent my first 36 years being bombarded by stimulus from all sides, at every moment of the day. My relationships with women up until then were meaningless connections. My first attempt at marriage drained me of my life force. Mykonos was my chance to empty the hurt and start again. I was a pharmacist. Pharmacists made a professional wage in California. I had made the money honestly or at the Blackjack table. I had enough money for years on Mykonos. My problem was my need to be needed. I seriously needed to serve a woman. That is how I was trained. Alone, I was frozen with want.

Don't get me wrong. I truly enjoyed not being attached to my old, regular way of life when I was working. I found something that Milan Kundera wrote in *The Unbearable Lightness of Being*.

> *Two people in love, alone, isolated*
> *from the world, that's beautiful*

Heather and I found respite with each other. I didn't know then quite how damaged I was. I suspect that she had some demons also, but we wanted nothing from each other. We held each other and we slept together and we spent every moment together for a week or ten days. I am embarrassed. I only think that her name was Heather. I honestly cannot remember for sure. When she was gone, she was gone. A sweet memory with no regrets. We did not share addresses. She gave me a Roman Catholic Saint medal that I misplaced years ago. I gave her a Twenty five dollar chips from Harrah's Lake Tahoe that I had been carrying in my pocket..

I took her down to the quay one morning and held her hand as she climbed down into the small boat that would take her out to the ship. It was a sunny morning in

December. Probably in the 70s. I was wearing a red, long sleeve cotton Tee shirt and a wool Greek fisherman's cap. Heather was wearing a jacket and had a bandanna covering her red hair. I didn't cry, but she did and it was like a stake in my heart. I wanted to leap into the small boat and to beg her to stay with me, but I didn't. The fist in my abdomen gripped my insides and squeezed.

I walked to a small taverna overlooking the harbor. I could watch as the ship sailed out of sight. I ordered a glass of ouzo and made it milky with some water. The bartender brought me some slices of sausage, some wedges of cheese and some crusted white bread. I sat there for awhile, feeling empty. Then a person I knew came into the bar. Andrew, an archeologist from New Zealand. Andrew was working on the sacred island of Delos. You could see the peaks of the island in the distance, from the Mykonos harbor.

Andrew and I became drinking buddies every night, at the taverna. An ouzo hangover is memorable, my friends. You wake up with an insane thirst. The first long drink of ice water somehow makes you drunk all over again.

Those nights of ouzo fog did not last very long. The lightness was so excruciatingly sharp that I had to become engaged with American life gain. There rarely were Americans on the daily ship. When there were, they were the quintessential tourists. They wanted Greek dancers and young, sexy widows dressed in black from *Zorba The Greek*.

They always asked me what I did all day on Mykonos. They asked me if I was a writer. I told them that I went to the chapel every day, waited for the ship and drank ouzo at night. They looked at me funny and that made me more resolute. I was going to stay on Mykonos forever.

But, eventually, the lightness got to me and I returned to California because I was in love. I had fallen in love with a three year old girl named Christy. I married her mother so I could be her step-father. I raised Christy as if she was my own child and life was no longer light. I was a Dad and I had responsibilities so I got my ass back behind a pharmacy counter. I worked part time only until the gambling money ran out. I took a full time job again, in 1981. I was a Manager again by 1982.

Christy and Jim around 1978

Back in the Saddle
The Modern Era of Pharmacy Began
for me around 1982

The non-pharmacist store manager at the Pay 'n Save in Pittsburg, California was the best manager I have ever worked with. He recognized that I, as the pharmacy manager, was making a lot of profit for the company. I received a great bonus every quarter and the manager got the biggest bonuses of his career on the mainland. Much of it on my back. He left me alone and I loved that job. We ran a cult of personality, promoted generic dispensing and made a consistent gross profit of 40%. That is not a typo. 40%! Ron was accustomed to running his store as he saw fit. He had managed the Pay 'n Save on Kauai in Hawaii and was virtually a free agent. He never looked too hard at the way I managed the pharmacy.

I loved my job. We introduced a new pharmacy computer system and that made it easier to make money. The computer, as it made our job easier, became the singular most damaging influence on the job of working in a pharmacy.

Later on, I am going to be very critical of the way pharmacists are treated, the ever watchful computer productivity programs and pharmacy's darkest age.

However, let me present a proviso. The department you work in must make money. If it doesn't, you will be blamed. That is always the way it works. If some idiot middle to upper management MBA person signs ridiculous contracts with PBMs just to assure that your department stays competitive, you will be blamed in the end if there is not a satisfying return on investment. They will say that your wait time is too long. They will say that you do not respond to the drive through as fast as you should. They will blame you for too much inventory. You don't answer the phone fast enough. Shit flow downhill, even in pharmacy.

It will always be your fault and it will always be a variation of the same tune that is played in the veneer all day, every day. It will be harder than hell to both practice pharmacy and satisfy the spying MBA people who know absolutely nothing about what it is to practice pharmacy. You, my friends, have been relegated to fast food status in much of your work. That is the veneer. It won't be easy thrive as a medical professional in the current environment, but you have to figure out a way to do it.

You are a Doctor

You can start with your degree. You are a Doctor of Pharmacy. You invested six years of your life to get the degree. It cost you tens of thousands of dollars, much of it in student debt, and now some yayhoo tells you that you can't call yourself *Doctor*.

Nurses and Physical Therapists can do more with a Ph.D. than you can with your Pharm D. Their doctor degrees can help them land top administrative jobs in hospitals, for example. There aren't that many jobs like that in pharmacy, certainly not in retail.

As more medical professionals earn the honorific *doctor,* physicians are fighting back as if they are in a war of sorts. Many medical doctors are suspicious that once thousands of nurses and pharmacists have doctorates, they will go to the state legislatures for more authority.

Doctor Roland Goertz, the board chairman of the American Academy of Family Physicians, says that physicians are worried about losing control over *doctor.* It is a word that has defined their profession for centuries. If other disciplines use the honorific *doctor,* physicians fear that the loss of control over the profession itself will follow. He doesn't want patients to be confused.

If you go into the pharmacy and the girl at the counter has *doctor* on her nametag, how many people are going to be confused when right below is the word *pharmacist?*

There was a time, about 120 years ago, when medical doctors' education was of questionable merit. Medical students paid the faculty directly. Can you imagine a student paying much if his grade was a "D"?

From the Encyclopedia of Chicago

In nineteenth-century Chicago, a medical degree was not always needed to practice medicine. No licensing laws yet governed medical practice, and doctors commonly learned medicine by apprenticeship or by reading medical texts. Doctors who had obtained formal medical training in eastern medical colleges founded similar schools in their newly adopted city. These schools enabled local youth to afford medical education and provided founders and faculty with income from students' fees while enhancing their prestige and reputations, which helped them attract paying patients.

In 1890, no state required a license to practice as a pharmacist. However, there were numerous colleges of pharmacy. The University of Texas School of Pharmacy was established in 1893. The school graduated its first women in 1897.

Pharmacy was more lucrative than practicing medicine. Doctors were basically surgeons and as late as 1918 (The *Spanish Flu* pandemic) were bleeding patients with influenza and some believed that the disease was caused by humors (putrid smells).

John M. Barry wrote *The Great Influenza*. It is a marvelously entertaining and informative book. In the early pages, Barry presents the argument that pharmacists were better educated than doctors and more respected around the turn of the century. That is not a stretch, in my opinion. The cocaine that the pharmacist could sell you did a wonderful job with symptoms and was preferable to the doctor's knife.

I have watched medical doctors and their strong lobbies protect their turf for decades and it is not always in the best interests of the patient. When there was a push by pharmacists to have Monistat Vaginal Cream put into a new behind-the-counter class of drugs, the medical doctor's establishment fought like hell to prevent it and this was only about protecting their turf.

The product ended up over-the-counter and can be legally sold in a truck stop. The doctors concern that the pharmacist might get a leg up caused thousands of women to listen to their Aunt Betty falsely diagnose their conditions as a yeast infection when it was bacterial. How many progressed to Pelvic Inflammatory disease?

A 2011 bill proposed in the New York legislature would bar nurses from advertising themselves as *doctors,* no matter their degree. Laws already in effect in Arizona, Delaware and other states forbid pharmacists and others to use the title *doctor* unless they immediately identify their profession.

In the early 1990s, I hired Cheryl Marinakis to work in the pharmacy at Pay 'n Save on Whidbey Island.

Cheryl is a Pharm D and insisted that *doctor* appear on her nametag. The non-pharmacist store manager had a fit.

"She's not a doctor," he argued. "She's a pharmacist. Why does she want to have *doctor* on her nametag?"

"Because she is a doctor."

"You are not a doctor. She'll make you look bad."

This guy had as a mission to make me look bad. "She's not a surgeon or a psychiatrist or a dentist, but she is a doctor, A Doctor of Pharmacy."

Cheryl won, of course. She dug in her heels and got her nametag. It was interesting that women patients asked for her after the new nametag when they did not before.

Think about this: A heavy-set guy wearing a wife-beater shirt and carrying a 24 can case of 20 ounce cans of Rainier beer calls out.

"Hey you, where's the lawn chairs for eight ninety nine?" He has bellies up to the pharmacy counter, blocking the way for prescription patients.

You ignore him, at first, but when he says, "Are you deaf or something. I aksed you where are the lawn chairs."

You finally walk over, "Sir, they can help you with the lawn chairs at the front of the store."

"You don't know? You work here, don't you?"

"I have enough to think about in the pharmacy department. I don't keep track of the lawn chairs."

"Are you getting smart with me, Missy?" He steps back and looks you over, his gaze lingering on your body. It makes you uncomfortable. He demands. "What's your name?"

"You can call me Doctor Simpson." You look at his face. It is red and puffy. This man wants to cause trouble.

"You ain't a doctor, Missy. What's your real name?"

"I am a doctor, Sir, a Doctor of Pharmacy." You do not intend to argue with this lout. "You do not need to know my name. When you make a complaint about me not knowing where the lawn chairs are, just say Doctor Simpson."

Your lead technician is a 20 something male, a big young man who is taking courses at the community college with aspirations to go to pharmacy school. He walks over and asks you, "Is there a problem, Brenda?"

"Brenda." The lout chuckles. "Why are you afraid of me knowin' your name?"

"She's Doctor Simpson to you, buddy."

The technician starts to leave the pharmacy, but you do not want a confrontation. You ask him, "Do you know where the lawn chairs are, Tony?"

That may seem as if it is a fantasy scenario, but don't be so sure. As long as you allow a jerk like this to call you by your first name, you are giving away some power. If you are a doctor, flaunt it. Show off. Request the honorific *Doctor* on your nametag.

The store manager in Cheryl's story was an abusive male. He was eventually escorted from the store by loss prevention because of sexual harassment. He never bothered Doctor Marinakis. Would he have taken liberties if she had been just Cheryl?

I don't see the retail pharmacist getting much advantage from the Pharm D degree. At least not right now. We all make the same money and we are all expected to be good managers of the Prescription Mill. We do not get paid for our clinical expertise presently, but that will change. Patients are smart rats and they will see just one more service that they can get for free at the drug store. When they notice that some are more equal than others (My apology to George Orwell, Animal Farm) in regards to their clinical expertise, the game will really change like a lightning strike.

Will the best clinicians make more money? Will the best clinicians be able to pick and choose where they will practice? I believe that the answer is affirmative. I can see the best looking, best educated, best prepared pharmacists making a lot more money than the

Prescription Mill managers. Unfair? Not at all. I believe that the days of every pharmacist making the same wage are numbered. Welcome to the real world.

Your professional presentation, all of it, will contribute to how much money you make. This is not your clinical expertise. It is your panache, your style, your flair. There are always ways to make your look appealing. Your professional presentation includes your personal appearance. Your clothes. Do you dress in a professional manner? Is your personality appealing? Have you developed your people skills? If you can say YES to all of these and have clinical proficiency, you are on your way.

I do see enormous advantages to the patients having a clinically trained pharmacist involved with their drug therapy. The problem is that the conditions that prevail in retail settings cause the average pharmacist, no matter what degree, to go numb and dumb down to just to get through the day. That is unfortunate. The best of you are left dying on the vine.

Doctors tend to spin every story for their own benefit. They argue that advanced degrees are just another step toward independent practice. For pharmacy, the Pharm D degree simply ensures that pharmacists stay competent in an increasingly complex drug therapy landscape. It's not like there is a conspiracy to over-educate pharmacists so they diagnose and prescribe.

If you have a Pharm D degree (most of you do), show it off. You earned it. It cost a lot in time, effort and resources. You are the expert in drug therapy. Put *doctor* on your nametag. The word *doctor* has magical powers. It engenders respect whether the holder deserves it or not. Semantically, this is a powerful psychological phenomenon. Take advantage of it. Use it.

There is a doctor in the east who said, "If it ain't broke, why fix it?" I think that you will agree that it has been broken for quite awhile. Medical doctors make mistakes. They can make minor errors and major errors and that is one good reason why you should refer to yourself as *doctor*. So you will have standing when you go to fix it.

Show some patience

Lately, there have been pharmacists who just do not think before they jump. They want all or nothing and when they go for it all, they end up with nothing. They are very righteous. For years, they tolerate conditions that other professionals would laugh at and then, all of a sudden, they rebel in one day and lose their job.

Take baby steps. You must counsel. The law says so, your personal standards say so and professional ethics say so. My advice is to pick and choose your counseling opportunities carefully. You will never have the time to give an eight point counseling session on every single new prescription. Trying to do that would confuse and bore the patient. Most patients will not listen to you for more than 30 seconds. Give them the highlights, ask if there are questions.

Your counseling and my counseling will be different on the same drug, but both will comply with the law.

Nobody can tell you how to counsel, just as no one can instruct the surgeon how to remove a tumor. I would bet that there is a drug store company that thinks that they can tell you the correct way to counsel and what to say on every drug. That is idiocy and it is an insult. No non-pharmacist, especially, can make any comment on the way you counsel and if a pharmacist thinks he can criticize your counseling, he is completely out of touch with pharmacy practice.

I counsel in depth when a woman is taking metronidazole. I invite her to the private counseling window and I ask her if it is for trichamonas vaginalis. If they answer that it is, I carefully counsel that they were most likely infected by their partner who is asymptomatic.

Trichomonas vaginalis is a sexually transmitted disease, but I can't say that. I simply say that their partner is asymptomatic and has no idea that he is infected. I have had to say, "No, your husband is not fooling around." I feel an obligation to counsel them in detail because otherwise they would be cured and re-infected immediately.

I have had to talk women down. They are ready to go kill their spouse at worst. The say things like, "I knew it"

I have to tell them to relax. "Your husband has not cheated on you. He doesn't even know that he has this bug. He needs to take this medicine too. Talk to your doctor about it. Otherwise, you will be infected again."

Those righteous pharmacists who just do not know how to take it slowly end up getting fired or they get reassigned. When pharmacists insist on practicing pharmacy and neglect the metrics, CVS will as a matter of policy put them on the floater team. They have no regular store and have to drive, sometimes long distances, to get to work. This is a punishment although I have had pharmacists report that their life is much better on the float team.

A woman in Tennessee is proceeding systematically. She has printed up stickers to put on the prescription bag stating that federal and state laws require the pharmacist to counsel. She uses these so that the technicians can't say, "Oh, I didn't know you had to counsel on this prescription."

She has been documenting everything for months. She has retained an attorney just in case CVS tries to get draconian with her. They moved her to the floaters. She expressed relief, but I wonder how she will feel when they assign her to work in a busy store a hundred miles from her home.

I corresponded with a new graduate a few years ago. She lamented the fact that she did not counsel the way she had been taught at school. I suggested that she go slow, but she was young and strident. She was righteous.

She was going to force the drug store company to make accommodations for her to counsel. She had the law on her side or so she thought.

Don't be an idiot and ruin your good thing by reacting and putting your job in danger. The *good thing* I refer to is a job close to home where you have terrific benefits and have earned multiple vacation weeks. You can take action, but slowly. Watch your back. Do not give them anything to jump you over.

Do not trust your technicians. Some may be trustworthy, but not all. I worked with a Washington State Level A Pharmacy Technician. It was not a slam dunk to get a Level A license in the early 1990s. The technicians had to qualify.

I trained this woman. I administered her qualifying examination. I treated her well. As long as I was earning a nice bonus, I gave her and the other two Level A Technicians a one hundred dollar bill at Christmas time. This Technician and I worked together on Saturdays. I picked her up on my way to work and took her back home after our shift. She stabbed me in the back. I believe that she felt it was better to ally with the non-pharmacist store manager.

There was a dispute with an unruly customer. I invited the guy to leave and never come back. The next day, the store manager calls me to the office. I was being written up over the incident. The Technician was the only witness. The manager didn't even try to hide the fact that she had busted me. Be very careful about trusting your Technicians.

The fantasy of the non-pharmacist store manager in Washington

"I've always been an advocate of strict discipline, Miss Frenzy!"

Maybe a spanking would be better than the dreaded write-up. It only would hurt for a little while. The write-up stays in your permanent record. The write up seems to be a modern management tool. In order to get what they want, CVS and Rite-Aid use the threat of the write up to influence your behavior.

"If your metrics are consistently in the red, Harriet, I will have no choice but to write you up." With the implication that after so many times in the penalty box, you will be let go.

The write-up is the whip or paddle or the electric prod of modern drug store management. I have made plenty of mistakes during my career. But I had never been written up until I started having difficulties with non-pharmacist management. I did not recognize the authority of non-pharmacist managers and told them so.

Never sign the write up. If you feel pressured, sign it only after you have written this on the signature page: "I am signing this under duress. I do not agree with anything written on this document" or something similar. If at all possible, bring a witness with you to the write-up *interview.*

My friend Ronald Benson (a CVS pharmacy manager in Alabama) gives very good advice. *If there is a dispute, never touch anyone, never use bad language and never ever walk off the job.*

The Dark Era of Pharmacy

The dark clouds began to gather when the first MBA mid-level management type went to the executive suite and gushed, "We can watch over these guys. We can watch their every move. All we have to do is write the programs."

Of course, he was talking about spying on the pharmacists. It did not take long before *wait times* became more important than providing professional services. I am not going to even pretend that counseling has ever been a priority to most pharmacists. But, for the purposes of argument, let's pretend that pharmacists actually consider counseling to be a vital part of the job. How can you satisfy the metrics and still counsel appropriately?

103

We are struggling with an outmoded paradigm. Pharmacists no longer have to be the prescription-fillers. The sooner we get Advanced Technicians the better.

Do you refuse simple compounding because you know that if you walk away from "The Mill" for just ten minutes you could be written up for being behind for the next two hours?

Do you ignore the patients who ask for your help with Over the Counter items because you can't take the time to give them your assistance?

Before the computer productivity programs were first written, pharmacists had the freedom to use their discretion and walk away from The Prescription Mill when they felt that it was important to be one on one with the patient.

I do not have to be laborious about the Dark Era. Every single one of you knows exactly what you put up with. They have you isolated out on the edge, away from the pack, and they can do anything they want to you.

If pharmacists ever needed to talk to each other, the time is now. But, we don't. There was a time when we had local pharmacist groups that met periodically for meals and conversation. I remember many pleasant meals with other pharmacists. It was usually a very early, before the stores opened, breakfast. If we had problems at work, we could ask the advice of our colleagues and friends. That went away years ago when the drug store chains became enemies. That attitude of distrust and rancor filtered down from the executive suite. Pharmacists became enemies too.

The stance that we are competitors rather than colleagues is never more apparent than when we call for a transfer. We keep each other on hold. We are curt and unhelpful. We act as if the other pharmacist is trying to take food out of the mouths of our children.

Come on, you guys. Isn't the job hard enough? Do we have to make it more uncomfortable? Take the chance to actually talk for a moment with the other pharmacist. There is a human being over there. You will find out that she has the same concerns you do.

Occupy

Pharmacy

I'LL BELIEVE CORPORATIONS ARE PEOPLE... WHEN TEXAS EXECUTES ONE!

First, be very clear that pharmacists have been protesting for two decades. I honestly am not sure that they would label their behavior a protest, but think about it. Pharmacists whine and complain. They say that they hate the job. They go about undermining pharmacy. They go out of their way to be sure that their children go to school to become teachers. Anything but pharmacy. The blame the profession when what they complain about is the job. Specifically, the veneer of the job.

They don't complain about the practice of pharmacy. If they do, they should be teachers. As I see it, there is unfocused carping because they see an amorphous beast in the shadows that they can't quite identify.

My intention in writing *Occupy Pharmacy* is to give that monster some form so that we clearly identify what we are up against. Once we understand what represents the real danger to us, we can fix it, I believe. We have one distinct advantage. They can't even open the pharmacy door without a pharmacist present. What about that power can't you complainers understand? I understand that Pennsylvania allows technicians to sell prescriptions while the pharmacist is out of the department for uninterrupted meal breaks. I am uncomfortable with that. Is this a violation of OBRA 90? I suspect that Rite-Aid (a Pennsylvania company) may be behind it.

It has been known for years that pharmacist executives from large companies have infiltrated the state boards. This is not right. It must change. It is like your husband coming home from Vegas with a hickey on his neck and saying. "Nothing happened, honey."

YOU HAVE THE RIGHT TO REMAIN SILENT AND LET THE1R MONEY SPEAK FOR YOU

(OR YOU CAN DEMAND TO BE HEARD)

Occupy Wall Street and *Occupy Pharmacy* may seem to be 180 degrees apart, but they are not that far away from each other. The main difference is the people represented by the protests. *Occupy Wall Street* has spread like a dust storm all over the planet. From *Occupy Honolulu* to *Occupy*

Galveston to *Occupy London*. *Occupy Madrid* saw huge crowds. The people represented are liberals, Retirees, Tea Party type, Libertarians, Teachers, Independents, Nurses, Anarchists, Teamsters, Socialists, Maids, Communists, Environmentalists, Indian chiefs and others.

The main issue people have with the *Occupy* protests is that they do not have a goal. That is not the case. There are many distractions like: what kind of coffee they will buy. A five minute "Let's have coffee" talk lasted hours when they argued about free trade coffee versus non free trade coffee. They have too many Indian chiefs.

Still, the *Occupy* protests have managed to define the argument. They are the 99% and they are protesting that the 1% has been gambling with the economy of the United States and they have rolled snake eyes. To make it worse, the 1% predicted that billions of dollars were going to be lost so they shorted their own investments and made money while losing money.

The bail outs were resented by the 99%. The banks privatized profits and socialized risk. When they won a bet, they kept the winnings. When they lost (and there was a lot of losing) they managed to get the citizens of the country to cover their bad bets.

Students in the *Occupy* movements are protesting the conditions that contribute to their being unable to find work. They have crushing student debt and cannot find a job in the field that they invested all of that money and time to earn the skill. They are young and they are angry. Some of them have to live with their parents at age 26. It is a sad commentary on the culture that has rewarded hard work and education for centuries. New pharmacists cannot find jobs in retail. Some of these pharmacists graduated from one of the new for-profit pharmacy schools. They do not know how to compound. They have limited retail experience. They are not smart rats. Of course, they will have difficulty.

My younger step son, Cody, is in law school, finally. A few years ago, he earned his Bachelor of Arts degree in Philosophy from the University of Montana. He quickly found out that his degree was, for the most part, worthless as far as helping him get a job. It was, however, his ticket into the game. My older step son did not get a ticket into America's game. He is working as a contractor in Afghanistan. He lives apart from his wife and four year old son for ten months of the year.

Cody has been married to a young woman who is a Doctor of Osteopathy. They have lived apart for most of their marriage because Cody had to fall back on his default job skill. He has made money as a cowboy since he was ten years old. The last cowboy job he had before law school was birthing calves during the spring season. It was a wet, dirty job and he was on call 24 hours a day.

There are thousands (if not millions) of young people with degrees who do not have a default job skill. They are waiting tables, selling cars, playing video games. They are frustrated and *Occupy* is ready-made for them.

Young people are only one group. Their youth makes them energetic. Older people may not be able to camp out, but they seem to be coming in waves. Nurses, teachers and other professionals. There are truck drivers and farmers showing up.

They agree on one thing. The corporations and the banks are raping us. They do nothing, but move money. Medical insurance companies and the Pharmacy Benefits Managers are *Occupy pharmacy's* example of the banks.

Occupy Pharmacy has no distractions. Our protest is diverse, but that array is restricted to *age, gender and race*. We are not all young, white women. We are, however, all pharmacists. Our fundamental complaint is that pharmacy has been taken out of the hands of pharmacists and has been bastardized by non-pharmacist MBA managers. Although these managers do not make the final decisions, they are the bright stars at the huge drug store company corporate offices and they have the ears of the executives. Pharmacists usually know better, but the Chief Executive Officers don't listen to us.

What has been the results of not listening to the pharmacists (or a committee of the brightest pharmacists)? It is pretty ugly. I learned never to call people insulting names, but I am going to. You will know who I am talking about when I point at *the idiots* who keep on signing those ridiculous PBM contracts. What are they thinking? They leave pharmacists twisting in the wind, hanging from a short strangling noose.

Then, when the profits aren't up to snuff, they take away hours from the pharmacy budget. You end up with no technician on duty after 8:00 PM. The pharmacist is expected to work the first morning shift alone. Pharmacist overlap is eliminated because some MBA numbers cruncher doesn't understand that pharmacists are medical professionals and that pharmacy is different than the camera department.

I strongly suggest that the executives look at where the problem originates when there is not a satisfying profit in the pharmacy department. The *idiots* who approve of those ridiculous PBM contracts are to blame, not the people working at breakneck speed in the pharmacy.

Pharmacists have been the scapegoats for years. They seem to think because they pay us a professional's wage that they can work us over whenever there is a dip in the bottom line. Sorry, we are on to you, big fellows. It's not our doing and we know it.

My message to the executives of the chain drug store companies is, "Shame on you. You need us more than we need you. A rush of pharmacists to the Small Business Administration applying for start-up loans to open their own pharmacy would just mean that you will have to pay the remaining pharmacists even more just to keep your pharmacies open. Worse than that, the people you have to pay so much won't be the brightest knives in the drawer."

"So, knock it off. Start showing some respect for the professional staff before it is too late. You are not the only games in town."

I have noticed that CVS is getting a lot of attention with *replace the pharmacist we have with a pharmacist we don't have*. The shortage may be over in many urban areas, but the people you have in place are better than the people you do not have. They have been loyal for years. You think that they are interchangeable with a new graduate from one of the for-profit boutique schools? That is dreaming. I suggest that you slow down and be very careful. I can actually foresee incidences where a pharmacist may actually buy a Rite-Aid's prescription files rather than the other way around. How long can that company stay in a death spiral with a stock price50 cents less than what they charge for a 20 ounce Diet Coke.

You have allowed the lemming-like MBAs to lead you to the edge of the cliff and soon there will be no turning back. This kind of questioning will label you as not being a team player. Pharmacists have been questioning for years. Nobody with power listens.

The *Occupy Wall Street* movement protests the centralization of power and money. There are too few people running our system. They are paying themselves enormous bonuses whether the gambles they take win or lose. The protestors demand that they quit gambling and cheating. They want the big banks to cease the gaming that has come so close to bankrupting the entire western world. The protestors want regulations back in place. They want the players to be responsible.

Occupy Pharmacy knows that centralization of power and money has gotten us into this mess. Pharmacy is a profession and pharmacists have discretion to act in a wide array of professional behaviors. With the corporations so top heavy with power and money, it is actually a maldistribution of power that we are faced with. So much power in so few hands has proven to be toxic. Our industry has been poisoned by incompetence and mediocrity.

We need Smarter, Better Business People Doing the Jobs that Make a Difference

What the non-pharmacist MBA Masters of the Universe have been doing has frozen everything and we cannot move on. We have been stuck for two decades.

They continue to do the same thing over and over again and they expect a different result. Perseveration. It is a authentic mental illness.

Very little truly new has happened in the pharmacy industry for a very long time. Walgreens says what we want to hear, but is *Power* just a way to eliminate high paying jobs as some pharmacists fear? If Walgreens is truly genuine, how will this play in Peoria? How long will it take for the new Walgreens to filter down to the stores? Walgreens is the biggest drug store chain with over 8,000 pharmacies. Does the company have a viable trimtab to get this huge ship on a new course? I will give the powers at Walgreens a free pass, for now.

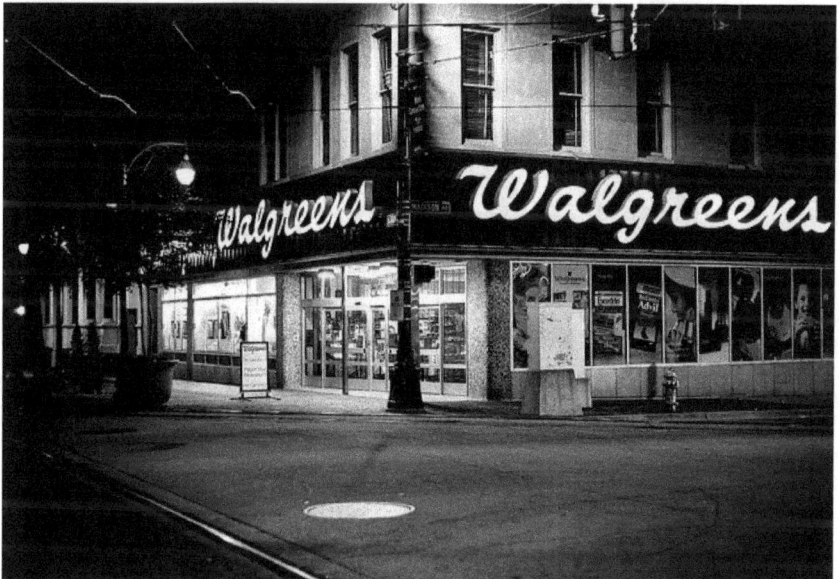

Rite-Aid has a reputation for throwing money to solve its problems. The 15 minute guarantee was just an old program dressed up with gift cards. Nothing new and they actually expect to get more business. Rite-Aid continues to do the same things. They believe that if they do it better or harder they will succeed. I have news for them. They failed a long time ago. Rite-Aid reminds me of a 1950 horror movie of the walking dead. The company is "The Big Zombie". Rite-Aid is dead, they just don't know it yet. The powers at Rite-Aid need to fall on their swords. They have been using fear to kill off good pharmacists for years.

CVS-Caremark is "The Big Evil". If you take a few minutes and Google "CVS in trouble" you will find pages and pages of lawsuits, civil actions, anti-trust investigations and consortiums of local independent pharmacists who have banded together in an attempt to make CVS play fair. It seems to me that the powers to be at CVS operate in a manner that takes into account the fines that they will have to pay as a necessary business expense. CVS agreed to pay a $77 million fine for failing to legally record pseudoephedrine sales. The operative word is *agreed*.

CVS has been using fear as a business tool. They manage pharmacists in a manner that many of the best, most loyal employees are afraid of losing their jobs. This may be a business strategy, but it is not how a profession is run. They treat the professional staff as if they were workers making blenders. I cannot imagine that middle management pharmacists came up with this business strategy on their own. Non-pharmacists are the drivers of this fear-based strategy.

Occupy Wall Street is a motley crew of independents, liberals, anarchists, socialists, libertarians, progressives, tea party, environmentalists and homeless people. It is a wonder that they have been able to focus on their message and have it be appealing to the millions of Americans who have been watching. The corporations that are their target have been waiting for the Occupy movement to fall apart. Instead, it has hardened and taken a form that people can watch. Police brutality has helped the *Occupy* cause.

Occupy Pharmacy is a movement among pharmacists. It includes pharmacists of all ages, men and women. All races are represented. Occupy Wall Street has many different sectors. Occupy Pharmacy has one. Corporate brutality has helped the *Occupy Pharmacy* cause.

Critics have accused *Occupy Wall Street* of a lack of structure. That is not, however, a lack of focus. It is a lack of trust.

Occupy Pharmacy has an eagle eye fine focus. We have watched our profession as it has been sucked dry. We started out in the 1950s as a 40% profit prescription business. In the 21st Century, I don't think we are even as good as a 10% profit prescription business. This degradation could not have happened without people in power in pharmacy being complicit.

When I was in pharmacy school in the early 1960s, the questions was: *Would you ever work for Revco?* Revco was an unfinished pine plank shelving deep discounter. In the Midwest, Revco signaled the end of pharmacy being a 40%

profit prescription business. The town where I practiced pharmacy for a brief time had three drug stores. A very classy small pharmacy, a big store out of town that featured a lunch counter and a downtown drug store that included a cosmetic department that presented expensive department store fragrances such as Chanel, Dior and Dana's Ambush. The men's section highlighted Caswell-Massey, English Leather and Dana's Canoe among other lines.

Revco came to town and within a year the out of town drug store and the small pharmacy had closed. Revco did not compete with the downtown store in the cosmetic department. The Revco lines were Maybelline and Cover Girl. The downtown store lost prescriptions by the handfuls, but the cosmetic department kept the store open. Forty years ago, this owner knew the importance of a niche business if he was going to stay in business. He knew that he could not compete with Revco with prescription prices and stay alive. He lowered his prescription prices just to keep his client base loyal. He expected them to buy his other products and they did. The cosmetic department was where he made his living.

For me, Revco represented the end of pharmacy as I knew it. The standard before Revco was customer service. After Revco arrived, the drum beat was price and only price. The chains became competitive. When PBMs entered the picture, the non-pharmacist MBAs seemed to forget that pharmacy is a profession because they caved in repeatedly. They allowed themselves to be bullied no matter how loud the pharmacists screamed.

Walgreens was the first (and so far only) company to refuse to sign ridiculous PBM contracts. They backed CVS-Caremark down and now (coming up in January, 2012) they are refusing to execute a contract with Express Scripts that does not include satisfactory reimbursements. It is about time. The PBMs are bullies, just as all of the medical care insurers.

Drug store corporations have been whining about declining revenues due to the recession in an industry that is damn near recession-proof if the company is managed properly. The real culprit, you see, is their gambling and fraud. The gambling is most noticeable with the PBM contracts they sign. If there is not enough profit, the MBA Masters of the Universe ferret out waste by claiming that the technician budget is too rich. They cut the technician hours and you know what happens. Customer service suffers and the pharmacist middle managers are supposed to fix that. Fraud is more difficult to pin down, but you only have to look at far as CVS. They are known for cutting corners. They play the game fast and loose and are under attack everywhere you look if you Google *CVS in trouble.*

The way this has affected pharmacists is that our working conditions have been degrading for forty years. Pharmacist have been sucked dry. We have been thrown under the bus. We have been raped to cover the bad gambling debts of the executives and the non-pharmacist MBA Masters of the Universe. They are bean-counters who look only at the projected profitability of their proposals. They are not pharmacists, however, and that is why they will not have consistent success. They are in the business of selling a product. They do not consider that pharmacy is a profession and that pharmacists are medical professionals.

When you have a CEO who graduated from Pepperdine University in Malibu, California (John Standley, Rite-Aid) and who was previously the CEO of Pathmark Stores, a regional grocery chain, you have a person who has no experience in pharmacy. He may have people to educate him about pharmacy, but that is like reading a book on sky-diving and thinking that you know what it is like to jump out of an airplane.

John Standley has no idea what it takes to get prescriptions out in fifteen minutes. He must actually believe that all Rite-Aid pharmacists have the time and the sales talents to enroll six patients a day to get flu shots. John Standley and the non-pharmacist MBA Masters of the Universe in Camp Hill, Pennsylvania have never worked a full day on their feet without a regular meal or rest period. They are very good at writing pharmacists up for not performing up to standards. They fire pharmacists on the slimmest evidence of cause.

We sit back and watch horrible things happen to good, loyal employees and we say nothing. We see, but we don't see. Some people just can't process the horror when a veteran pharmacist with a good record is let go because he takes the time to counsel patients and the metrics go red too often when he is on duty. This is what psychologists call *Normalcy Bias*. When we find ourselves in unsettling circumstances, we shut down and pretend everything is normal.

Many of us suffer from *Motivated Blindness*. We don't see what is not in our interest to see. In the late 1980s, I was working with an older pharmacist who was the PIC. I knew that the company wanted me to replace him as the Pharmacy Manager. I had been successful at a previous store and they trusted that I could duplicate that success. The company agreed to transfer me to Oak Harbor, Washington from Pittsburg, California. In some respects, I was a hired gun. I admit, I wanted this job.

This pharmacist was an old-fashioned druggist who was clearly not interested in what the company wanted him to do. He was a chatty Cathy with all of the old customers. He was a devout Mormon and had a bad habit of proselytizing at work. He was slow and set in his ways. When he made a serious dispensing error, I sat aside and looked the other way as the Pharmacy District Manager went after him. He had dispensed piroxicam (Feldene) when the prescription was clearly written for nifedipine (Procardia). A serious blunder. How could he make such a mistake? I was the one who discovered the error. When the patient with the pasty, grey-hued face asked me if the drug was right, I had to tell him the truth.

The Pharmacy District Manager could have just said, "You have been a loyal employee. Thank you for your service. We want you to step down and let Jim take over."

What he did was take the PIC up to the office for the *write up*. When the meeting was over, I was the new Pharmacy Manager. The pharmacist I replaced was stricken. He had been with the company for three decades. He had been loyal and he expected some loyalty in return. The last I heard, he had bought a pharmacy in Utah.

I have always liked to believe that I was a champion for the downtrodden. I was a man who would stand up for this pharmacist. He made an error. He had always shown competence previously. He was a pillar of the community. This was unjust, unfair and unwanted. I don't want the job under these circumstances. I wanted the job with honor. I thought that I would fight for this guy. I believed I was a fair man.

"He made a mistake. All of us have made dispensing errors. He has been a loyal employee for thirty years. You can't demote him." But I didn't say a word.

I'd bet that I would have taken a highly indignant position if the scenario had been presented to me as a hypothetical. I know that I would have taken the high road and protested what I thought was a travesty. In reality, my mind (according to a psychologist) somehow grasped what was going on and rushed a protective filter into place. I did not have to confront the horror of this demotion. It was not in my interests. It was *Motivated Blindness*.

Then there is the *Bystander Effect*. There were plenty of pharmacists watching what would happen. I can almost compare it to the apocryphal Kitty Genovese case in the 1960s. Bystanders stood and watched Kitty get beat to death on the sidewalk. This pharmacist was not beat physically, but psychically he was tortured. He had to wait weeks until the day came for the *write up*. He couldn't sleep. It was as if he knew what was coming. Every pharmacist in the district was watching.

We are really good at self-deception. We have blind spots. We pay attention to the facts we are comfortable with, the ones we like and keep the ones we don't like in check. We inflate our own merits and predict that we will behave better than we actually do. As Max Bazerman and Ann Tenbrunsel write in their book, "Blind Spots", *When it comes time to make a decision, our thoughts are dominated by thoughts of how we want to behave; thoughts of how we should behave disappear.* Pharmacists are quick to throw their colleagues under the bus. This is the underside of our nature. It is pathetic that professionals act like this, think like this and do not examine their behavior.

Do medical doctors, lawyers, engineers or accountants act like this? Do they behave like they are in a concentration camp and are liable to be gassed at any minute? I don't think so.

We are in a difficult situation in retail. The job has devolved over the decades since Durham-Humphrey. We are now often treated like members of a second, lower class . The culture of selling-a-product is to fault. The MBA Mastesr of the Universe are supposed to ferret out waste and impress the executives with their good ideas. They are not pharmacists, however, and they disregard that pharmacy is a profession. As long as we let them get away with it, nothing will change.

How do we overcome our natural tendency to evade and self-deceive? That was a proper question that I could not answer when I watched the pharmacist get demoted. Honestly, I did not even know what the question was. I do not expect you to answer that question. We will always evade the darkness in our souls and deny the underside of our own nature. It is important that we know what the question is.

In the beginning of the 1970s, we did not use computers. It was very difficult for our bosses to spy on us. We sent in our weekly reports on paper, but the value to the pharmacy district manager was questionable because he had to determine on his own what they meant. There was no computer to compile all of the numbers and come up with conclusions. In the 1970s there were really only two numbers that counted. Daily prescription count and profit. If you did enough prescriptions and sent enough money to the company's account, you were a favored employee. The chains exerted only a modest framework for the pharmacy managers to work on.

In 2012, the chain drug store culture tries to control everything. They now have sophisticated computer programs to spy on everyone and everything you do. If you dare to take five minutes to heat up the noodle soup that your spouse made for you to take to work, the metrics are negatively affected, the timers go red and you get a black mark at the district office.

Can you just see yourself, "But, Mister Tony, I was heating up the noodle soup that my husband made for me.

I didn't even get a chance to eat it. I let it go cold. I work my butt off for Big Evil and you won't even let me eat".

"Tough shit, Brenda," says your pharmacy district manager, the pharmacist Capo, "If you have to eat, we'll just find a pharmacist that doesn't have to eat." Later, they find him drunk in a strip club, with a black eye, blood on his shirt and missing teeth. He had been calling one of the strippers "Mommy" and said that he needed to be spanked.

The one thing that the company's can't control is your discretionary behavior. I am not talking about the veneer. I am talking about you practicing pharmacy. When you counter-prescribe, when you compound, when you counsel and follow all of the laws and regulations that govern pharmacy, when you live up to your own persona professional standards, you are untouchable.

This schematic is pharmacy only and does not include the rest of the store

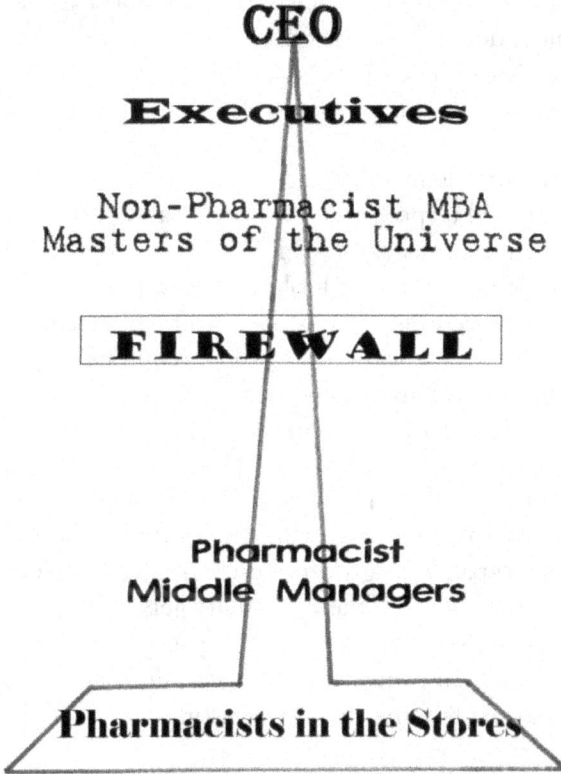

CEO

Executives

Non-Pharmacist MBA
Masters of the Universe

FIREWALL

Pharmacist
Middle Managers

Pharmacists in the Stores

Hierarchy
of Power and Responsibility as it is in
modern chain store pharmacy

Three men, the CEOs of Rite-Aid, CVS and Walgreens control the business of approximately 20,000 pharmacies in the United States. They also control the fate of more than 50,000 pharmacists. That is not scary of itself. Idiots do not become CEOs. What is frightening is that these CEOs depend on non-pharmacists to come up with the plans and programs that the CEOs will back.

The problem is that the Masters of the Universe have no cultural responsibility for pharmacy. You know that a fifteen minute guarantee is a recipe for disaster, but the MBA looks at it like he looks at selling panty hose. MBAs have three jobs. Find waste. (Cut technician hours). They search for new ways to grow like requiring the pharmacists to give six flu shouts a day or else. This is business. It is not a profession.

There is a firewall between the Masters of the Universe and the pharmacists and all influence flows down hill. Smart, capable drug store merchant pharmacists have good ideas, but none of their creativity gets by the firewall. Pharmacists must assert their authority as practicing professionals. They must learn to view themselves as active participants rather than spectators.

Occupy Pharmacy is not here to bring the chain drug store corporations to their knees. We are here to improve their sense of responsibility. We cannot continue to give them a free pass. It is our profession. Of course, they are in business and they have to make money, but pharmacy is a profession and they must respect that. We don't want anarchy, we want fundamental change in the paradigm of management.

This schematic is pharmacy only and does not include the rest of the store

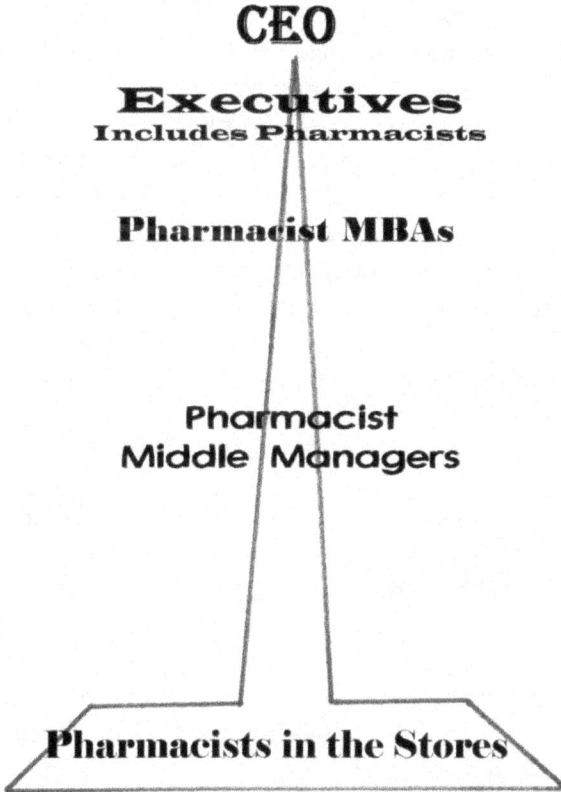

CEO

Executives
Includes Pharmacists

Pharmacist MBAs

Pharmacist
Middle Managers

Pharmacists in the Stores

Hierarchy
**of Power and Responsibility as it MUST Be
if our profession is to thrive in the JOB of
working in a chain drug store**

This schematic speaks for itself. The line of creativity and responsibility moves downward and upward with everyone in the company contributing and taking responsibility. When I was a pharmacy manager in the early 1990s, the company had a committee of working-in-the-stores pharmacists that met monthly with a committee of the company's executives. The pharmacists signed off on just about every program that we took on. Granted, this was prior to the pervasive productivity programs. I still sent in reports on paper every week. This was a relatively small chain, just a few hundred stores in the Pacific Northwest. I don't remember ever being led by the nose by an MBA Master of the Universe. It was a simpler time in the 1990s. We ran pharmacies and pharmacists were left alone to practice as we saw fit. We were respected and we responded with respect back.

Some drug store corporations are like the big banks. The banks gambled on subprime mortgages and engaged in fraud. The drug store corporations engaged in rampant, uncontrolled discounting beginning forty years ago. They gambled and they lost big time with the PBMs. Every year, the margins got leaner and every year the idiots kept on signing the ridiculous contracts. This is a cultural death wish and you know who pays the price, don't you? The workers in the pharmacy. The pharmacists and the technicians. Our jobs have become a miserable shadow of what they once were.

Banks have gotten away with bad debts, usurious credit card agreements and with skimming huge profits on the backs of people like you and me.

Chain drug store companies have gotten away with wild growth selling cheap and disposable goods and giving away profits from the professional department. They have done it on the backs of people like you and me. This model is not sustainable. They cannot continue to keep on doing the same things, just more and better, again and again because they will fail. In the end, they are the same tired old programs just dressed up.

We have a crisis in pharmacy. The corporations control most of the money and money is power. The issue is that there isn't that much excess money anymore. 90% of all prescriptions are paid by Medicaid or the PBMs. Do you remember when they threw money at you? I knew a woman named Rajinder who worked for Rite-Aid in a town near the Canadian border in the Pacific Northwest. Rite-Aid was pharmacist poor. They often called Rajinder on Friday and practically begged her to work an extra shift on Saturday. She asked for quadruple pay and she got it. Those were the days. There was plenty of money and it all went straight up. Everybody was happy. They treated pharmacists as highly educated medical professionals. There were no wise guy Masters of the Universe making horrible moves and shirking the responsibility.

I know that you are afraid that you will lose your job if you do not do a good impersonation of Step-n-Fetchit. There are suddenly plenty of pharmacists, but, trust me, there are not that many true merchant pharmacists. If you learned the drug store business and not just how to manage in the veneer, you will do fine. If you are young and have not grown up in the drug store culture, learn it. Your life depends on it.

In the end, they are afraid of you. Your pharmacy district managers are terrified that they will lose *their* jobs. If enough of you in one district quit caving and toeing the mark, they will be done, back in a store and lucky to be the pharmacy manager. The Masters of the Universe are too ignorant to be afraid. They do not really have a clue about the drug store business. They are numbers-crunchers. It pisses me off mightily that the executives have so disrespected our profession.

Corporate control of pharmacy has led to deep systemic problems and I don't know if they can be solved without blowing it up. Can we demand that state legislatures institute regulations like we can with the out-of-control banker? It is up to us, the practitioners, to demand regulations. Corporate control has become like in-breeding. The results have not been pretty. I have come to the grim realization that the corporations have rigged the system and they can't even see the trouble they are causing.

All they need to do is settle down. Demand satisfying contracts with the PBMs or no contract. Treat pharmacists well and pharmacists will perform, just like the olden days. A happy pharmacist will make money.

We are suffering death by a thousand small cuts. Every pharmacist has her own cuts to report. They killed off a twenty year career Rite-Aid pharmacist because she is needle-phobic. A psychiatrist stated that her condition is a real disability, but that didn't matter to Rite-Aid. They killed her off and denied severance benefits. A CVS pharmacist was fired when he determined that the lack of help and one prescription every 80 seconds was dangerous. He closed the drive through and lost his job. There are too many stories of lives shattered when all they wanted to do was the right thing.

Pharmacy's Most Valuable Resource

They can't even open the pharmacy if you are not there. They can't sell one prescription if you are not there. There is no other person who can legally counsel. There is no one other than a pharmacist who even knows how to compound. Unfortunately, some of the newer for-profit pharmacy schools are not even teaching compounding, but that is a story for another time. The pharmacist is the person who has to check each prescription for accuracy. The pharmacist is the one responsible if there is a mistake. The pharmacy accounts for up to 70% and more of the dollar volume of the entire store. What about that kind of power is so had to understand? You always have the chance to take your talents elsewhere and even go and open your own pharmacy. If you stay in the neighborhood, you could be competing with inexperienced pharmacists who can barely paint by the numbers.

You are important and still they have treated you with crushing disrespect for the last 25 years or more. You have been living in psychological prison. A metaphorical Stockholm Syndrome. At its worst, you think it is normal.

I was a houseguest in the home of a couple who both are school psychologists. The three of us were sitting at the breakfast table. They had been asking me about pharmacy. The husband thought that I had it pretty good. He was marveling at how I had been able to move around in my career. At that point, it was in the late 1980s. I had worked in three states and never had a problem getting a job for top wage. I was feeling pretty smug until the wife did something that astounded me.

She took a sip of her coffee and went to the counter. She picked up the phone and said eight words that shook my world.

"This is Carol. I won't be in today."

What the? How can a professional person make a call like that? Didn't she have a boss? Didn't she have work to do? From that day on, I have looked at how we are treated with a tight jaw, teeth grinding. We are the cash cow and they have a habit of treating us like pieces of pharmacist shit. Can't they see that we can bust it up for them at any turn?

Once more, I will be redundant and repeat that is it their own fault if the pharmacies don't make enough profit. They were idiots playing the discount game starting in the early 1970s. They were idiots playing the dollars off coupon game. They were idiots playing the gift card game for anything including looking sideways at a difficult customer. They have been idiots for signing those ridiculous PBM contracts. They got themselves into a damn big hole and they want us to stand on our shoulders to get out.

CVS and Rite-Aid seem to be dedicated to riding herd on the backs of the pharmacists, spurs digging, whips in hand. They keep on doing more of the same. They are one trick ponies. Walgreens, I am watching with ever increasing interest. I don't think they will be hiring any paint by the numbers pharmacists. If they are serious about patient-centric pharmacy they need people who can actually do it.

The centralization of power that has been made possible by computerization has gotten us in this mess. We have a maldistribution of power. People who have proven to be very poor stewards of the profession have had way too much power. MBA Masters of the Universe have the ear of the executives while pharmacists throw up their hands and say, "This is a frikkin' terrible idea" when they see what is expected of them.

There was a time when the pharmacy manager and his/her staff handled the issues and problems that came up on a daily basis. The problem would be identified. The staff would react to the problem and they would implement the solution.

The pharmacy I managed in 1992 had a nursing home contract. Part of the job was a monthly inspection of the facility. I was supposed to inspect the storage of the drugs, inventory the narcotic locker and verify that there were valid orders. The biggest part of the job was to inspect the drug therapy of individual patients. Then, one afternoon, I was called to the administrator's office. He was sort of smug when he told me that if I wanted to keep the contract, I would be expected to make regular visits to interview patients for the "self-medication" program. I knew that the local independent wanted this contract.

I did not hesitate. This would take me away from the pharmacy three more afternoons a week for an hour, but I agreed immediately. I set up the weekly schedule and never called my pharmacy district manager for approval. This contract was a money maker. The non-pharmacist store manager had a hissy fit. He whined because I would have to go to the nursing home three afternoons every week. I basically told him to suck eggs. He huffed and puffed and bypassed my boss and went to the district manager. I won the battle and the war.

Do that today? Forget it. In almost all situations where there is an issue, the pharmacist is simply a clog in the wheel. For long distance, the corporate types define the problem, the reaction and the ultimate solution. What a ridiculous way to run a business. They treat the pharmacy manager like she is a pharma-robot. Don't think. Do not make decisions. Just follow the program and don't get creative.

Remember the veneer? The corporations fear losing control when the only thing they control is the veneer. They have absolutely no control over your practicing pharmacy. The practice of pharmacy is a discretionary act and it is all yours. You may stand in front of a computer screen all day and feel like a cheap imitation of a professional, but you can walk away any time you want. Just do it a few times a day in the beginning. It is addictive. You will feel so terrific when you help an elderly patient understand her drugs and why she needs to be compliant that you will want more. That is called practicing your profession. Attending to the veneer is just something you have to do because we deal with a product and the game is still to sell more, unfortunately, for now.

We have a nationwide financial crisis in pharmacy because of the PBMs. Express Scripts claims they are "Charting the Future of Pharmacy." That statement epitomizes narcissistic arrogance. CVS and Rite-Aid are going along with that idea by turning their pharmacies into dispensaries.

Greed is the reason for every season in the 21st Century. This is exemplified by the huge banks, insurance companies and where it cuts close to the bone for us, the PBMs. Express Scripts and the rest are for-profit engines. They provide no health care. More that 20% of our total medical care expense of $2.8 trillion per year goes toward profiteering. That's a helluva lot of profit for companies that do nothing by move money.

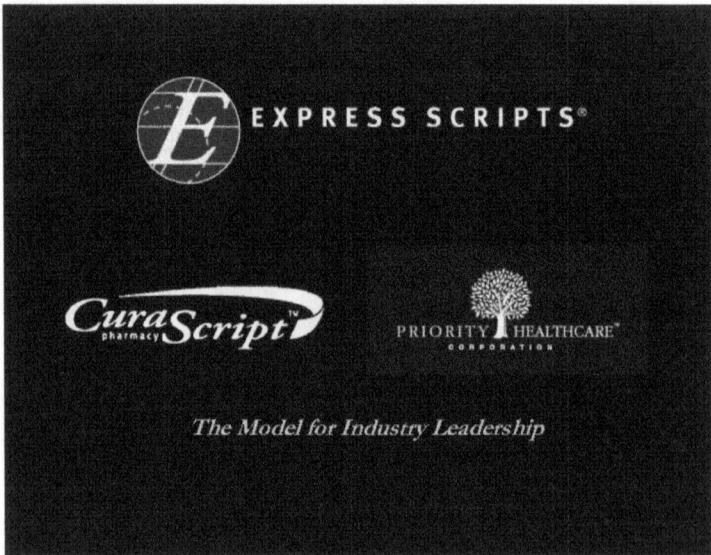

When we enter the civilized world, the government will buy all the drugs and they will pay us a fair and reasonable fee for dispensing and counseling. Let the good times roll. Customer service will prevail. The chains may have to offer delivery just to keep up. Big Pharma will have to go back to providing new drugs for serious illnesses rather than me-too products and the medical insurers and PBMs will be out of the game. That model is sustainable. What we are tolerating now is not.

Do Business Differently

I barely have some hope. I am an optimist. I look forward to the days when the big drug store companies become open for a great idea from anyone at any time. They must see that creativity can come from any employee and not only from the bean counters. The MBA Masters of the Universe have shown that they are hardly creative at all regarding pharmacy. They don't know enough about pharmacy.

We need for talented, experienced pharmacists to make better choices about their jobs. We need for pharmacists to stop keeping their mouths shut. This is how an ordinary pharmacist can make a difference. Right now, there is no conversation. Speak up. We want all pharmacists talking about the condition of our job in retail chain pharmacy. The good jobs and the bad jobs. We need to identify which is which and to hold the bad employer's feet to the fire. CVS and Rite-Aid seem to be looking for ways to fire good pharmacists and to replace them with warm bodies. They must be made to pay. This is not a sustainable model. If they continue this way, they will lose big time and companies like Walgreens will rule.

For complete disclosure, I do work part time for Walgreens. I love what they are claiming they will do. It is a huge company with lots of nooks and crannies. I suspect that there are middle managers out there who do not like change so we'll see how this shakes out.

Anyone in the drug store industry who says he has no idea what the pharmacists are protesting is disingenuous at best and a liar at worst. The damage done by ignorance is glaring. This is not a rejection of capitalism, by any means. Capitalism is good. This is a rejection of unexamined non-pharmacist management gone haywire. We need oversight by pharmacists who will take stewardship of our profession. Some of these huge companies are like Elvis, gorging on banana splits because there is no one there to say, "No, you can't do that, Elvis."

Which of These do we Need in the Conversation?

Institutional Reform

Professional Awakening

UNION

I have never been a big fan of pharmacist unions. I still have the idea that we can create a win-win collaborative relationship with the corporations. I believe that a happy pharmacist will be a productive employee and that this will have a positive impact on the return on investment. But, what do I know? Unless the companies make a sea change in thinking and paradigm-busting, revolutionary changes, I don't see much hope. Unionizing may be the only hope. So bit it.

Unions can only exist in the status quo. Unions cannot exist in a vacuum. There has to be a malevolent and nasty management to go up against for there to be unions.

If you are going to conspire to get unionized, do it smart. Bring in the big guns. First, contact the Guild for Professional Pharmacists. They started in the 1970s in California when they organized Savon Drug Stores in the San Diego area. If you are going to do this, prepare for a battle. You can take a loan against your 401k for living expenses. Don't just walk out. You will lose your job. If you have to strike, make sure the action complies with the Department of Labor rules.

The State Boards are Part of the Problem

Pharmacists have been begging their state boards to intercede on their behalf for decades. The job of the boards is to regulate pharmacy and pharmacists to assure that the public is not harmed. It is not the job of the board to act to improve your working conditions. The Pharmacy Alliance has been promoting the idea that a tired pharmacist is a dangerous pharmacist. Here is a letter to the editor that appeared in the October, 2011 edition of Drug Topics magazine.

I was executive director of the NCBOP when the 12-hour rule was adopted and applied. This was a direct result of a dispensing error made at an Eckerd drugstore by a pharmacist working a 16-hour day with no meal break and/or scheduled bathroom break. Eckerd claimed that she "volunteered" for this shift.

We also implemented a policy, based on a Rule, which limited the number of prescriptions per day per pharmacist. I believe it would be in the interest of public safety for other states to take similar actions."

David R. Work
Executive Director Emeritus
NORTH CAROLINA BOARD OF PHARMACY

The proof is in the pudding. We will see how the North Carolina Board of Pharmacy acts in the Kelly Hoots case. Perhaps, there is hope. No word yet, as of this writing.

The boards of many states are controlled by chain store executives. There is no question that more mistakes are made by tired pharmacists at hour 12 of a 14 hour day. If the boards do not take action, we will have to go to the individual state legislatures. We'll have to push for legislation that will force the boards to act. Right now, the boards are protecting the corporations and not the public.

You do it, You own it.

There must be accountability for the mess that these MBA types and executives have gotten us into. They started trying to tweak a viable business model in the 1970s in order to compete; and like Topsy their ability to dig an even deeper hole just growed and growed and growed. They must be held accountable for the stupid choices they have made. They have taken a solid profession right to the edge with their gambling and throwing money at every problem.

Move Beyond Anger

It doesn't help at all to scream and holler and let our anger control our behavior. We are liable to make horrible choices when we act out. The outcomes are rarely desirable. Righteousness will win nothing. You make enemies at best and lose your job at worst.

I suggest in "The Prisoners of Comfort" that anger is your friend. It is a vital message and I will repeat it here.

Recovery is regaining Your Power

Essentially, the miserable pharmacist is wretched because they choose to be unhappy. There is a choice every single day to be proud of what they do or to blame the job because they are not happy. They don't even use the best tool available to them. That tool is anger!

Anger is fuel. It is not the bad thing that your parents said to suppress as mine did. "Jimmy, nobody needs to know you are angry. You should control yourself." We feel anger and we become frustrated when we hide it because we want to do something about it. This goes against the image of the calm, in-control professional. Instead of showing the anger, we stuff it and chug Maalox and take two 20mg omeprazole every day.

How would it look if we showed that we were angry? At work, you don't hit that someone or break that something or throw that fit. If you smash that fist against the wall, do it in the bathroom where no one can see that you are out of control.

What we do with our anger is deny it. We stuff it so far down that we forget what makes us angry. We are institutionalized and we believe that we should not get angry. We lie about being angry at the store manager. We hide our anger at the lack of technician help. We do not express our outrage to the district manager. Doesn't he know that it is his precious customer service that pays the price?

Some of us hide it so well that we medicate the anger and filch the occasional lorazepam to hide it even better. We are professionals and professionals are nice people. We bury our anger. We block it and we hide it.

What we do best with our anger is lie about it. Unfortunately for our spouses, we lie so well that we often take our misery out on the people we love (or are supposed to love) the most. We do everything but listen to our anger.

Listen to your anger. That is what it is meant for. Anger is not a polite request. Anger is a scream. It is a command. It is a slam of the fists down on the table demanding your attention. Anger has a right to be heard. Anger should be appreciated and valued. Anger must be listened to if you are to regain your professional balance and power. Why? Because anger is an atlas or a chart or a diagram back to living the ideals you had when you were in pharmacy school.

Anger reminds you of your boundaries and limits, the areas where no one was allowed to tread without your permission. If you can set up the periphery of your professionalism in just one area, more will follow. If you list only ten serious drugs that you will counsel on no matter what, your list will be twenty in little time. If you let the store manager know in writing that his touching you at anytime, in any manner, is unwanted, you will regain enormous power and control over your own life on the job. You can gain power simply by refusing to get wet underpants because you neglect going to the bathroom when *you have to go. Documenting anything at work that makes you uncomfortable will give you surprising control.*

Anger shows us where we want to go. We may not know exactly what we do want on the job, but our anger tells us, without ambiguity, what we sure as hell do not want. That is a really good place to start because anger shows us where we have been and sets us on the course of recovery. Anger is not a sign of disease. It is a sign of health. If you no longer get angry at being institutionalized, stop, take a deep breath, and examine how you will find your way back. I contend that you will find that the first sign of recovering your health, well-being and pride will be anger. Welcome it. Savor it.

It is not very healthy to act out from anger. That is childish and not productive. I quit a job once out of anger. It was a good job. I was well respected in the community. The problem was that the store manager tried to micro-manage my department. I have never bent to management from a non-pharmacist. This guy was out to bring me to my knees. I fell right into the trap. I became so angry that I brought the problem to a head with some stupid brinksmanship. My district manager did not back me as fully as I wanted, so I quit. My one-way commute for that job was less than ten minutes. The one-way commute for the next job was ninety minutes. I was like a teenager having a meltdown. I turned my anger into indignation without any examination of the circumstances. I was an idiot.

Anger is there to be acted upon. Anger points the direction. Anger is the wind for our sails as our sailing ship tacks as we move on the appropriate bearing where our anger guides us. Had I used my head and had the presence to translate what the anger was telling me, I would have made better choices.

"Damn it, I could run a better pharmacy than that!" This anger says that you want to have your own pharmacy, you just need to put all of the pieces together.

"I can't believe it. Mildred told me that she was going to demand a transfer to the suburbs and she got it. That's what I wanted." This anger says: Stop keeping your goals and dreams hidden. You need to express your wants and believe that you deserve your dreams to come true.

"That was my idea. This is unbelievable. I mentioned it only once and that son of a bitch took my plan and put it to work. He gets all of the credit and I get none." This anger says that it is time to take yourself seriously and show yourself some respect. Your ideas are good enough to do something about.

Anger is the tornado that blows away all of the restrictions and hesitations and lack of self confidence of our old lives. Anger is a valuable instrument to be used productively. Anger cannot be the master, only the servant. Anger is a deep well of power, if used properly.

Apathy, laziness, misery and gloom are the enemies. Anger is not a good buddy, but anger is a friend. Not a mild-mannered friend, but a very loyal and steadfast friend. Anger will always remind us when we have been cheated or cheated upon. It will always tell us when we have been deceived or when we have betrayed ourselves. Anger will tell us that it is time, finally, to act in our own best interests. Anger is not the action itself. It is the action's invitation.

Think before you act from anger. The satisfaction will be brief. The consequences may last a very long time.

A Message to the CEO
"You will be held responsible for EVERYTHING."

Larry Merlo, the CEO over at CVS-Caremark received $10,906,613.00 in compensation in 2010.

Absolutely Everything

When you are in a knife fight, bring all of your tricks. There are only a handful of people at the top in the drug store industry. We need to identify who has the power and then find the weakness and attack it. The glaring weakness, the most crippling weakness, and it has greatly harmed our industry, is that too many people with the power are not pharmacists. They may be good merchants in the grocery business or in big box stores, but they are not pharmacists. To compound the weakness, they go to non-pharmacists for guidance.

Before the Demands

There will never be a right time. The time will always be the wrong time for action. Stop complaining. That is just noise. It is time to practice pharmacy. If you meet resistance, visit an attorney and have a simple, but compelling letter written. Have it be a straightforward letter of intent. The letter should be sent to the upper management of your company with a copy to your state board of pharmacy.

Have the letter state that it is your intention to comply with all pharmacy laws, state and federal. You can leave it at that, but I would have the attorney list all of the laws that concern you, with the numbers. Then, watch your back. Document anything and everything that you perceive as retaliation. Transferring you to the float team is retaliation.

Life Is Too Short To Not Love Your Work

The Money Will Never Be Enough.

You must find fulfillment and Satisfaction

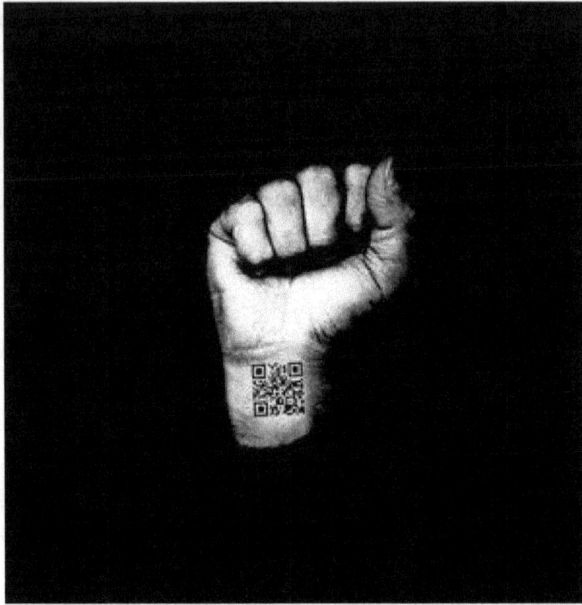

The Comfort Demands

This is the bang for the buck. Straightforward, unambiguous and uncomplicated. Put them on the front burner. If we can ever get these, watch the change.

Pharmacist will be managed by pharmacists. Non-pharmacists will have no authority over pharmacists. This is their own fault and they need to take responsibility for loose-cannon, non-pharmacist managers who are bullies.

For five decades, non-pharmacist managers have resented that pharmacists make more money than they do. They perceive that pharmacists do not work as hard as they do. That is so much tough shit. Pharmacists are well-educated highly-trained medical professionals. If the non-pharmacist store manager buys too many dozens of eggs for Easter, the back room may smell of hydrogen sulfide. If a pharmacist makes a mistake somebody can die. Only an idiot can fail to see the significance. We demand that non-pharmacist store managers have no authority over pharmacists.

Recently, an unreasonable customer called the store manager back to the pharmacy. It was a busy Friday evening and she was demanding that I call the doctor at UTMB for her father's prescription. I refused. The nurse had promised to send the Rx via E-Prescribe.

I told this woman, "Maam, we are going to let the nurse do her job. Your father's prescription will show up on our screen when it does. Then we will fill it. If I called, I would get the Access Center. We would not hear from the doctor's nurse until Monday at the earliest."

She turned to the store manager on duty and demanded, "Make him call the doctor."

The manager said, "I am not his boss, maam. I can't make him do anything."

The woman sputtered, "But, I don't have time to…"

"…Jim is the pharmacist on duty. He is the one in charge. It is his call."

Walgreens.

Pharmacists will be included in the decision-making process when new programs are implemented or existing programs are modified. A working in-the-store pharmacist or a committee of pharmacists will have veto power over the good ideas that the MBA Masters of the Universe come up with.

You can bet that no pharmacist was involved on Rite-Aid's fifteen minute guarantee program. CVS's new trend of finding ways to fire loyal veteran pharmacists and replace them with warm bodies who can barely paint by the numbers was not the idea of a pharmacist.

In the 1990s, Pay 'n Save had a committee of pharmacists that met regularly with a committee of executives. The Vice President in charge of pharmacy was a pharmacist. No new programs were implemented without the approval of the committee of pharmacists.

Granted, it was a simpler time, a down-to-earth period before the PBMs began to chart the downward profit spiral that, left unchecked, can doom us. I was sincere in my suggestion that the government buy all the drugs and pay us a reasonable fee for dispensing and counseling. Think about it. It would benefit you. If you have visions of being some kind of Ayn Rand captain of industry, give it up. It is a form of mental illness in the 21st Century.

All drug store companies will assign a working in-the-store pharmacist to act as an ombudsman to be responsible for investigating and resolving complaints from pharmacists against the company or anyone within the company. The ombudsman will have the ear of the CEO. It will be a full time job. The environment created will be vastly more creative than what they are doing now. The company's pharmacists will be back in the loop. We will have a win-win situation again after years of being treated as if we are piece work employees running dispensaries instead of pharmacies.

The pharmacy manager will have exclusive hiring and firing powers over all pharmacy personnel. No one will be allowed to do any work in the pharmacy until they have been vetted by the pharmacy manager.

How many times has the non-pharmacist manager sent a clerk from the camera department to work in the pharmacy without consulting with the pharmacy manager? This must cease.

Pharmacy Technicians must be well-trained and qualified. The C.Ph.T. designation must be a minimum qualification. Ideally, the individual states must, at the very least, limit the licensure of technicians to individuals who have the C.Ph.T. qualification. It would be best if the state boards institute a legitimate examination process for technicians. After working with many technicians, it is clear that the C.Ph.T. designation does not guarantee competency.

The Nuts and Bolts Demands

There will be a 30 minute uninterrupted rest/meal break for every 8 hours worked. There will be paid 15 minute breaks. One in an 8 hour shift and two in a shift of 12 hours. There will be an additional 30 minute rest/meal break in any shift longer than 12 hours. The second 30 minute break will be paid.

Pharmacists will be encouraged to use the bathroom when they have to go rather than waiting until it becomes critical. Pharmacists may take their time. They should be able to reapply makeup, adjust their clothing or look at themselves in the mirror. Pharmacists will never be made to feel guilty for using the bathroom. I am an older man who takes 25mg of hydrochlorothiazide every morning. I make frequent bathroom breaks in the morning. I try to go when I have to go, but so many decades of bad habits cause me to have near emergencies too often. Damp under garments are undignified. We should never tolerate circumstances that cause us to wet ourselves.

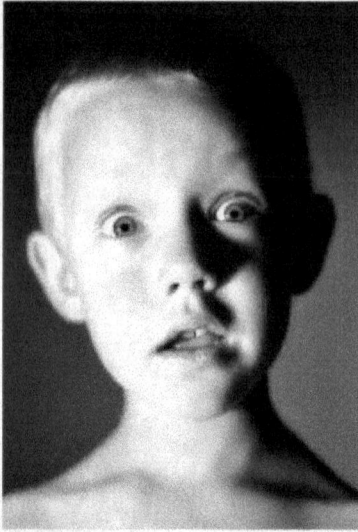

Let me get this straight: The company will side with petulant, unreasonable, angry, demanding customers instead of with me, its loyal employee?

And this is meant to lead to better customer service?

Customer complaints will be managed by the pharmacy manager with deference to the pharmacist. The pharmacist will not be blamed or assumed to be at fault until the pharmacist has had the opportunity to explain herself.

The Customer is Always Right

This has always been bullshit and it always will be bullshit. There are some pretty smart rat customers out there. I recently had two women from Houston present prescriptions from a pain clinic. They each had three prescriptions and they were the same. #240 Hydrocodone/APAP, #120 Alprazolam 2mg & #120 carisoprodol 350mg.

I declined to fill the prescriptions. There were plenty of audience members watching as they loudly demanded to know why. I told them that I did not have to explain myself. It was my discretion, my call entirely. They got nowhere with the manager on duty so they returned to the pharmacy and started ranting.

"You have to fill our prescriptions. It is the law."They had bullied their way to the front of the counter.

"Stop ignoring us. Why won't you fill these prescriptions we got at the doctor's?"

After a good, solid ten minutes of their badgering me, I had finally had enough. "The law is very clear, ladies. If the pharmacist does not believe that the prescriptions are not for a legitimate medical purpose, we are obliged to decline to fill them. I doubt that these are for a legitimate medical purpose. That's why I refuse to fill them. It is entirely my call. The law is on my side in this."

I spoke in a normal voice. I should have had them come to the counseling window where I could talk with them in privacy. On Monday, the store manager told me that they had called the district office and screamed HIPAA violation. Very smart rats.

I have worked for companies where I would have been written up for such a thing. I have worked with two non-pharmacist store managers who had their fondest dreams come true when they could jump me and threaten the dreaded write up.

What this company did was send a three page summary of the HIPAA regulations and had everyone who worked in the pharmacy sign it. That was okay with me. I let this one get out of control. I needed the reminder.

I demand that management at least act as if they believe that the pharmacist is innocent of any transgression when there is a complaint.

This would be nice. "Jim, a customer from Tuesday named Bill Jones claims that you were rude."

"I refused to fill his prescription for Vicodin because he had no refills. He wanted loaners and I refused that request too. If that is rude, then I was rude."

"I thought you guys gave loaners."

"Yes, for maintenance drugs, but never on controlled substances. At least, I won't."

"Okay cool. I thought it was something like that."

The Customer is Not always Right

The Keely Hoots Demand

The demand is that the operation of the Drive Through window be managed by the pharmacist on duty. Kelly Hoots closed the Drive Through on a day when his most competent technician was absent. He was working alone with one inexperienced technician and no cashier. They were filling prescriptions at a rate of one every 80 seconds. The two of them were expected to do everything, including manning the register and the Drive Through. Mister Hoots perceived that the situation was perilous and that patients were being put in danger. He was not able to consistently counsel appropriately. Kelly closed the Drive Through. He was fired by CVS.

The pharmacist on duty will always be authorized to close the Drive Through when she perceives that patient care is compromised, pharmaceutical care is not being delivered or a frenetic pace is endangering the patient.

I do not know if any of the 50 states in our country authorizse a non-pharmacists to have authority in the pharmacy. As far as I know, every state recognizes that the legal Pharmacist in Charge is the person with 100% power in the pharmacy. When the Pharmacist in Charge is absent, the Pharmacist on Duty is the person with all of the clout. The non-pharmacist store manager has absolutely no say-so in the pharmacy. Check the state laws where you work. Post them prominently. Document any and all incidences when a non-pharmacist usurps the pharmacist's authority. We must demand that the drug store companies make it very clear that the pharmacist is the last call.

In the Kelly Hoots vs. CVS case in North Carolina, it will be interesting to see how the Board of Pharmacy rules. The non-pharmacist store manager clearly usurped Kelly's authority when he came into the pharmacy and reopened the drive through. He can't do that. Legally, Kelly holds all of the cards.

We demand that drug store companies make unambiguous statements that non-pharmacists have no power in the pharmacy.

I am quite sure that there are plenty more demands that I can list here, but if we get the ones I have cataloged here, life at work would take a quantum leap from the soul-crushing working conditions to a sane, safe and satisfactory work experience with no stops in between. I have only one word if the companies ignore us.

UNION

OVERSIGHT

This is our profession. We have allowed non-pharmacists to mold our jobs. That has to stop. Pharmacists must maintain oversight over the MBA Masters of the Universe and even the CEO.

The free pass that we have given them just because of their positions has to stop. If the idea is stupid, we need to tell them. If the idea is dangerous, we need to shake them and tell them. If the idea is going to fail and result in a multi-million dollar loss, we need to grab them by the collar and leave some blood and teeth on the floor.

Start being the Hunter instead of the Hunted.

They are afraid of You

They are idiots for not treating you well and you are idiots for taking abuse when you hold all of the trump cards. They do not even have a business without you. With no pharmacy, Rite-Aid is a poor example of a variety store. What about that power do you not understand?

My last words today:

Practice Pharmacy

www.ingramcontent.com/pod-product-compliance
Lightning Source LLC
Chambersburg PA
CBHW061319220326
41599CB00026B/4951